Gang Tian

Canonical Metrics in Kähler Geometry

Notes taken by Meike Akveld

Birkhäuser Verlag
Basel · Boston · Berlin

Author's address:

Department of Mathematics
Massachusetts Institute of Technology 2-246
77 Massachusetts Avenue
Cambridge, MA 02139
USA

2000 Mathematical Subject Classification 35C55; 58G30; 81T30

A CIP catalogue record for this book is available from the
Library of Congress, Washington D.C., USA

Library of Congress Cataloging-in-Publication Data

Tian, G.
Canonical metrics in Kähler geometry / Gang Tian.
p. cm. -- (Lectures in mathematics ETH Zürich)
Includes bibliographical references and index.
ISBN 3-7643-6194-8 (alk. paper). -- ISBN 0-8176-6194-8 (alk. paper)
1. Kählerian manifolds. 2. Geometry, Differential. I. Title.
II. Series.
QA649.T46 2000
515'.73 – dc21

Deutsche Bibliothek Cataloging-in-Publication Data
Tian, Gang:
Canonical metrics in Kähler geometry / Gang Tian. - Basel ; Boston ;
Berlin : Birkhäuser, 2000
(Lectures in mathematics)
ISBN 3-7643-6194-8

ISBN 3-7643-6194-8 Birkhäuser Verlag, Basel – Boston – Berlin

Printed on acid-free paper produced from chlorine-free pulp. TCF ∞
Printed in Germany
ISBN 3-7643-6194-8

9 8 7 6 5 4 3 2 1

Lectures in Mathematics
ETH Zürich
Department of Mathematics
Research Institute of Mathematics

Managing Editor:
Michael Struwe

Contents

Preface

This monograph results from the author's lectures at the ETH during the Spring Semester of 1997, when he was presenting a Nachdiplom course on Kähler-Einstein metrics in complex differential geometry.

There has been fundamental progress in complex differential geometry in the last two decades. The uniformization theory of canonical Kähler metrics has been established in higher dimensions. Many applications have been found. One manifestation of this is the use of Calabi-Yau spaces in the superstring theory.

The aim of this monograph is to give an essentially self-contained introduction to the theory of canonical Kähler metrics on complex manifolds. It is also the author's hope to present the readers with some advanced topics in complex differential geometry which are hard to be found elsewhere. The topics include Calabi-Futaki invariants, Extremal Kähler metrics, the Calabi-Yau theorem on existence of Kähler Ricci-flat metrics, and recent progress on Kähler-Einstein metrics with positive scalar curvature. Applications of Kähler-Einstein metrics to the uniformization theory are also discussed.

Readers with a good general knowledge in differential geometry and partial differential equations should be able to understand the materials in this monograph,

I would like to thank the ETH for the opportunity to deliver the lectures in a very stimulating environment. In particular, I thank Meike Akveld for her patience and efficiency in taking notes of the lectures and producing the beautiful LaTeX file. Without her efforts, this monograph could never have been as it is now. I would also like to thank Ms. Nini Wong for her endless patience in proof-reading and correcting numerous typos in earlier versions of this monograph.

Part of my work involved in this monograph was supported by National Science Foundation Grants DMS-9303999 and DMS-9802479, at Courant Institute of Mathematical Sciences and Massachusetts Institute of Technology. My research was also supported by a Simons Chair Fund at Massachusetts Institute of Technology.

MIT, April 1999. Gang Tian

Chapter 1

Introduction to Kähler manifolds

1.1 Kähler metrics

Let M be a compact C^∞ manifold. A Riemannian metric g on M is a smooth section of $T^*M \otimes T^*M$ defining a positive definite symmetric bilinear form on T_xM for each $x \in M$. In local coordinates $x_1, \ldots, x_n$, one has a natural local basis $\frac{\partial}{\partial x_1}, \cdots, \frac{\partial}{\partial x_n}$ for TM, then g is represented by a smooth matrix-valued function $\{g_{ij}\}$, where

$$g_{ij} = g\left(\frac{\partial}{\partial x_i}, \frac{\partial}{\partial x_j}\right).$$

Note that $\{g_{ij}\}$ is positive definite. The pair (M, g) is usually called a Riemannian manifold.

Recall that an almost complex structure J on M is a bundle automorphism of the tangent bundle TM satisfying $J^2 = -\,\mathrm{id}$.

Definition 1.1 *The Nijenhuis tensor $N(J) : TM \times TM \to TM$ is given by*

$$N(v, w) = [v, w] + J[Jv, w] + J[v, Jw] - [Jv, Jw]$$

for v, w vector fields on M.

An almost complex structure J on M is called integrable if there is a holomorphic structure (that is a set of charts with holomorphic transition functions) such that J corresponds to the induced complex multiplication in $TM \times \mathbb{C}$. Clearly, any complex structure induces an integrable almost complex structure. The following theorem is due to Newlander and Nirenberg, see for example Appendix 8 in [14].

Theorem 1.2 *An almost complex structure is integrable if and only if*

$$N(J) = 0.$$

The hard part is to prove that $N(J) = 0$ implies integrability.

We say that J is compatible with the metric g or g is a Hermitian metric if

$$g(u,v) = g(Ju, Jv).$$

We can then define a 2-form ω_g by

$$\omega_g(u,v) = -g(u, Jv).$$

Usually, we call such an ω_g the Kähler form of g.

Example 1.3 Consider $M = \mathbb{C}^n$ as a real manifold by identifying $\mathbb{R}^{2n}$ with $\mathbb{C}^n$ in the usual way, then the corresponding almost complex structure J is given by

$$J\frac{\partial}{\partial x_i} = \frac{\partial}{\partial x_{n+i}} \text{ and } J\frac{\partial}{\partial x_{n+i}} = -\frac{\partial}{\partial x_i}.$$

Let $z_1, \ldots, z_n$ be the canonical complex coordinates, $z_j = x_i + \sqrt{-1}x_{n+j}$. Then

$$dz_j = dx_j + \sqrt{-1}dx_{n+j}, \quad d\bar{z}_j = dx_j - \sqrt{-1}dx_{n+j}$$

and

$$\frac{\partial}{\partial z_j} = \frac{1}{2}\Big(\frac{\partial}{\partial x_j} - \sqrt{-1}\frac{\partial}{\partial x_{n+j}}\Big), \quad \frac{\partial}{\partial \bar{z}_j} = \frac{1}{2}\Big(\frac{\partial}{\partial x_j} + \sqrt{-1}\frac{\partial}{\partial x_{n+j}}\Big)$$

for $j = 1, \ldots, n$. We also have that

$$J\Big(\frac{\partial}{\partial z_j}\Big) = \sqrt{-1}\frac{\partial}{\partial z_j} \text{ and } J\Big(\frac{\partial}{\partial \bar{z}_j}\Big) = -\sqrt{-1}\frac{\partial}{\partial \bar{z}_j}.$$

If g is the Euclidean metric, then its Kähler form is given by

$$\omega_g = \frac{\sqrt{-1}}{2}\sum_i dz_i \wedge d\bar{z}_i = \sum_i dx_i \wedge dx_{n+i}.$$

We will denote by ∇ the Levi-Civita connection of g, which is the unique torsion free connection which makes g parallel.

Definition 1.4 *A Kähler manifold (M, g, J) is a Riemannian manifold (M, g) together with a compatible almost complex structure J, such that $\nabla J = 0$, where ∇ is the Levi-Civita connection of g.*

Let (M, g, J) be a Kähler manifold. We can extend the metric g $\mathbb{C}$-linearly to $TM \otimes \mathbb{C}$. Recalling that $T^{1,0}M$ and $T^{0,1}M$ are the $\pm i$ -eigenspaces of J, we see that $g(u,v) = 0$ for $u, v \in T^{1,0}M$ or $u, v \in T^{0,1}M$ (use compatibility of J with the metric). Define $h(u,v) = g(u,\bar{v})$ for $u, v \in T^{1,0}M$, then this defines a Hermitian inner product on $T^{1,0}M$.

If (M, J) is a complex manifold and J is compatible with the metric, then from the above, we get that

$$g\left(\frac{\partial}{\partial z_i}, \frac{\partial}{\partial z_j}\right) = g\left(\frac{\partial}{\partial \bar{z}_i}, \frac{\partial}{\partial \bar{z}_j}\right) = 0$$

and in local coordinates, we can write

$$g_{i\bar{j}} = g\left(\frac{\partial}{\partial z_i}, \frac{\partial}{\partial \bar{z}_j}\right),$$

so we have

$$\omega_g = \frac{\sqrt{-1}}{2}\sum_{i,j} g_{i\bar{j}}dz_i \wedge d\bar{z}_j.$$

On a Kähler manifold M, a Kähler metric is uniquely determined by its Kähler form. So we often denote a Kähler metric g by its Kähler form ω_g. Note that $N(J) = 0$ on M, so M is a complex manifold and $d\omega_g = 0$, that is, ω_g is a closed form.

Proposition 1.5 *If the Nijenhuis tensor $N(J)$ vanishes, then $\nabla J = 0$ if and only if $d\omega_g = 0$. In particular, if (M, J) is a complex manifold and g is a Hermitian metric, then (M, g, J) is Kähler if $d\omega_g = 0$.*

Proof. For all $u, v, w \in TM$, we have

$$\begin{aligned} d\omega_g(u,v,w) =& u(\omega_g(v,w)) + v(\omega_g(w,u)) + w(\omega_g(u,v)) - \\ & \omega_g([u,v],w) + \omega_g([w,u],v) - \omega_g([v,w],u). \end{aligned}$$

Since $\omega_g(u,v) = g(Ju,v)$ for any u, v, we deduce from the above

$$d\omega_g(u,v,w) = g((\nabla_u J)v, w) + g((\nabla_v J)w, u) + g((\nabla_w J)u, v).$$

Replacing u by Ju or v by Jv in the above, we obtain

$$\begin{aligned} d\omega_g(Ju,v,w) &= g((\nabla_{Ju} J)v, w) + g((\nabla_v J)w, Ju) + g((\nabla_w J)Ju, v), \\ d\omega_g(u,Jv,w) &= g((\nabla_u J)Jv, w) + g((\nabla_{Jv} J)w, u) + g((\nabla_w J)u, Jv). \end{aligned}$$

Summing up the above two and using the facts that

$$J^2 = -I \quad \text{and} \quad g(Ju,v) + g(u,Jv) = 0,$$

we have

$$\begin{aligned} & d\omega_g(Ju,v,w) + d\omega_g(u,Jv,w) \\ =& 2g((\nabla_w J)u, Jv) + g((\nabla_u J)Jv - (\nabla_v J)Ju + (\nabla_{Ju} J)v - (\nabla_{Jv} J)u, w) \\ =& 2g((\nabla_w J)u, Jv) - g(N_J(u,v), w). \end{aligned}$$

It follows that if $d\omega_g = 0$ and $N_J = 0$, then $\nabla J = 0$. It is trivial to see that $\nabla J = 0$ implies $d\omega_g = 0$. □

1.2 Curvature of Kähler metrics

Let (M, g, J) be a Kähler manifold and ∇ be its Levi-Civita connection. We extend ∇ in a $\mathbb{C}$-linear way to $\Gamma(T_{\mathbb{C}}M)$.

Since M is also a complex manifold, we have local coordinates $(z_1, \ldots, z_n)$ and hence a local basis $(\frac{\partial}{\partial z_1}, \cdots, \frac{\partial}{\partial z_n})$ for $T_{\mathbb{C}}M$. We define the Christoffel symbols Γ_{ij}^k by

$$\nabla_{\frac{\partial}{\partial z_i}} \frac{\partial}{\partial z_j} = \Gamma_{ij}^k \frac{\partial}{\partial z_k} + \Gamma_{ij}^{\bar{k}} \frac{\partial}{\partial \bar{z}_k}$$

and

$$\nabla_{\frac{\partial}{\partial z_i}} \frac{\partial}{\partial \bar{z}_j} = \Gamma_{i\bar{j}}^k \frac{\partial}{\partial z_k} + \Gamma_{i\bar{j}}^{\bar{k}} \frac{\partial}{\partial \bar{z}_k}.$$

Because $\nabla J = 0$ and $J\frac{\partial}{\partial z_i} = \sqrt{-1}\frac{\partial}{\partial z_i}$ and $J\frac{\partial}{\partial \bar{z}_i} = -\sqrt{-1}\frac{\partial}{\partial \bar{z}_i}$, we see that

$$\nabla_{\frac{\partial}{\partial z_i}} \left(J\frac{\partial}{\partial z_j}\right) = J\nabla_{\frac{\partial}{\partial z_i}} \frac{\partial}{\partial z_j} + \left(\nabla_{\frac{\partial}{\partial z_i}} J\right) \frac{\partial}{\partial z_j}$$

implies

$$\sqrt{-1}\left(\Gamma_{ij}^k \frac{\partial}{\partial z_k} + \Gamma_{ij}^{\bar{k}} \frac{\partial}{\partial \bar{z}_k}\right) = J\left(\Gamma_{ij}^k \frac{\partial}{\partial z_k} + \Gamma_{ij}^{\bar{k}} \frac{\partial}{\partial \bar{z}_k}\right) = \sqrt{-1}\left(\Gamma_{ij}^k \frac{\partial}{\partial z_k} - \Gamma_{ij}^{\bar{k}} \frac{\partial}{\partial \bar{z}_k}\right)$$

and therefore $\Gamma_{ij}^{\bar{k}} = 0$, similarly, $\Gamma_{i\bar{j}}^{\bar{k}} = \Gamma_{i\bar{j}}^k = 0$, so the only possible non-zero terms are Γ_{ij}^k and $\Gamma_{\bar{i}\bar{j}}^{\bar{k}} = \overline{\Gamma_{ij}^k}$.

Moreover, if $g_{j\bar{k}} = g(\frac{\partial}{\partial z_j}, \frac{\partial}{\partial \bar{z}_k})$ denote the metric tensor in local coordinates, then

$$\frac{\partial g_{j\bar{k}}}{\partial z_i} = \frac{\partial}{\partial z_i} g\left(\frac{\partial}{\partial z_j}, \frac{\partial}{\partial \bar{z}_k}\right) = g\left(\nabla_{\frac{\partial}{\partial z_i}} \frac{\partial}{\partial z_j}, \frac{\partial}{\partial \bar{z}_k}\right) = g\left(\Gamma_{ij}^l \frac{\partial}{\partial z_l}, \frac{\partial}{\partial \bar{z}_k}\right) = \Gamma_{ij}^l g_{l\bar{k}}$$

and hence

$$\Gamma_{ij}^l = g^{l\bar{k}} \frac{\partial g_{j\bar{k}}}{\partial z_i} = g^{l\bar{k}} \frac{\partial g_{i\bar{k}}}{\partial z_j}.$$

So if the Kähler metric is given by $\{g_{i\bar{j}}\}$, its connection ∇ is given by

$$\left\{\Gamma_{jk}^i = g^{i\bar{l}} \frac{\partial g_{j\bar{l}}}{\partial z_k}\right\}.$$

Proposition 1.6 *(Normal coordinates in Kähler case) Let M be a Kähler manifold with a real analytic Kähler metric. Given $x \in M$, there exist local complex coordinates $(z_1, \ldots, z_n)$ unique modulo unitary linear transformations such that $g_{i\bar{j}}(x) = \delta_{ij}$, $dg_{i\bar{j}}(x) = 0$ and $\frac{\partial^l g_{i\bar{j}}}{\partial z_1^{i_1} \ldots \partial z_n^{i_n}}(x) = 0$ for all $l \geq 0$ and $i_1 + \cdots + i_n = l$, and this also holds for its conjugate.*

This proposition can either be proved by induction in an elementary way or for another elegant proof, see [16], page 286.

So given a Kähler manifold (M, g) with its Kähler form

$$\omega_g = \frac{\sqrt{-1}}{2}\sum_{i,j} g_{i\bar{j}} dz_i \wedge d\bar{z}_j$$

and its compatible connection ∇, the Riemannian curvature tensor is defined by

$$R(u,v)w = \nabla_u \nabla_v w - \nabla_v \nabla_u w - \nabla_{[u,v]} w$$

and can be extended in a $\mathbb{C}$-linear way to $T_{\mathbb{C}}M$.

Note that because J is parallel, that is $\nabla J = 0$, we have that

$$R(u,v)Jw = JR(u,v)w.$$

Defining

$$R(u,v,w,x) = g(R(u,v)x, w),$$

we can easily see that

$$R(u,v,Jw,Jx) = R(u,v,w,x)$$

and because of the splitting $T_{\mathbb{C}}M = T^{1,0}M \oplus T^{0,1}M$ into the $\pm i$-eigenspaces of J, we can deduce that $R(u,v,w,x) = 0$ unless w and x are of different type.

In local coordinates $z_1, \ldots, z_n$, this means that the only possibly non-vanishing terms are essentially

$$R_{i\bar{j}k\bar{l}} = R\left(\frac{\partial}{\partial z_i}, \frac{\partial}{\partial \bar{z}_j}, \frac{\partial}{\partial z_k}, \frac{\partial}{\partial \bar{z}_l}\right).$$

Using that

$$\nabla_{\frac{\partial}{\partial z_i}} \frac{\partial}{\partial z_j} = g^{k\bar{l}} \frac{\partial g_{i\bar{l}}}{\partial z_j} \frac{\partial}{\partial z_k},$$

we can deduce that

$$R_{i\bar{j}k\bar{l}} = -\frac{\partial^2 g_{i\bar{j}}}{\partial z_k \partial \bar{z}_l} + g^{s\bar{t}} \frac{\partial g_{s\bar{j}}}{\partial z_k} \frac{\partial g_{i\bar{t}}}{\partial \bar{z}_l}.$$

We define the Ricci curvature to be the trace of this, so we get

$$\mathrm{Ric}_{k\bar{l}} = R_{k\bar{l}} = g^{i\bar{j}} R_{i\bar{j}k\bar{l}} = -\frac{\partial^2}{\partial z_k \partial \bar{z}_l}(\log \det g_{i\bar{j}}).$$

We will denote the Ricci curvature both by Ric and by R_{kl} which should not cause any confusion. In complex coordinates we have found a nice expression for the Ricci curvature, but we need to check that it is still the same as that in the Riemannian case. So choose an orthonormal basis $e_1, \ldots, e_{2n}$ such that $Je_i = e_{n+i}$ for $i = 1, \ldots, n$ and set $u_i = \frac{1}{\sqrt{2}}(e_i - \sqrt{-1}Je_i)$, then $\{u_i\}$ is a unitary basis. It follows that

$$\begin{aligned}
R(u_i, \bar{u}_i) &= \sum_j R(u_i, \bar{u}_i, u_j, \bar{u}_j) \\
&= \sum_j \frac{1}{2} R(e_i - \sqrt{-1}Je_i, e_i + \sqrt{-1}Je_i, u_j, \bar{u}_j) \\
&= \sum_j \sqrt{-1} R(e_i, Je_i, u_j, \bar{u}_j) \\
&= -\sum_j R(e_i, Je_i, e_j, Je_j) \\
&= \sum_j R(Je_i, e_j, e_i, Je_j) + \sum_j R(e_j, e_i, Je_i, Je_j) \\
&= \sum_j R(e_i, Je_j, Je_j, e_i) + \sum_j R(e_i, e_j, e_j, e_i) \\
&= \sum_j R(e_i, e_{n+j}, e_{n+j}, e_i) + \sum_j R(e_i, e_j, e_j, e_i) \\
&= \mathrm{Ric}(e_i, e_i).
\end{aligned}$$

Here we have used the first Bianchi identity for R:

$$R(e_i, Je_i, e_j, Je_j) + R(Je_i, e_j, e_i, Je_j) + R(e_j, e_i, Je_i, Je_j) = 0.$$

This shows that the Ricci curvature defined above is the same as the one in Riemannian geometry.

Recall that if $|x| = |y| = 1$ and x is perpendicular to y, then $R(x, y, y, x)$ is the sectional curvature of the plane spanned by x, y. Set now

$$u = \frac{1}{\sqrt{2}}(x - \sqrt{-1}Jx) \text{ and } v = \frac{1}{\sqrt{2}}(y - \sqrt{-1}Jy),$$

then

Definition 1.7 *The bisectional curvature is defined to be*

$$R(u, \bar{u}, v, \bar{v}) = R(x, y, y, x) + R(x, Jy, Jy, x).$$

Definition 1.8 *A Kähler manifold (M, g) is said to be of constant bisectional curvature if there exists a constant λ such that in any local coordinates of M,*

$$R_{i\bar{j}k\bar{l}} = \lambda(g_{i\bar{j}}g_{k\bar{l}} + g_{i\bar{l}}g_{k\bar{j}}).$$

Example 1.9 Let $M = \mathbb{C}P^n = \{[z_0 : \cdots : z_n]; 0 \neq (z_0 : \cdots : z_n) \in \mathbb{C}^{n+1}\}$ and let $U_0 = \{[1, z_1 : \cdots : z_n]\} \simeq \mathbb{C}^n$ be an open subset of $\mathbb{C}P^n$. Set

$$g_{i\bar{j}} = \frac{\partial^2}{\partial z_i \partial \bar{z}_j} \log(1 + |z_1|^2 + \cdots + |z_n|^2)$$

or equivalently

$$\omega_g = \frac{\sqrt{-1}}{2} \partial\bar{\partial} \log(1 + |z_1|^2 + \cdots + |z_n|^2) = \frac{\sqrt{-1}}{2} \left(\frac{dz_i \wedge d\bar{z_i}}{1 + |z|^2} - \frac{\bar{z}_i dz_i \wedge z_j d\bar{z_j}}{(1 + |z|^2)^2} \right).$$

We will check that ω_g is globally defined on $\mathbb{C}P^n$. Let

$$U_1 = \{(w_0, 1, w_2, \ldots, w_n)\} \subset \mathbb{C}P^n$$

and check what happens to ω_g under a coordinate transformation on the overlap

$$U_0 \cap U_1 = \{[1 : z_1 : \cdots : z_n] = [w_0 : 1 : w_2 : \cdots : w_n]\}.$$

There exists a non-zero constant λ such that $\lambda w_0 = 1$, $\lambda = z_1$ and $\lambda w_i = z_i$ for $i = 2, \ldots, n$, so $\lambda = \frac{1}{w_0}$ and therefore $z_i = \frac{w_i}{w_0}$ for all $i \neq 1$ and $z_1 = \frac{1}{w_0}$. So we see that

$$\begin{aligned}
\omega_g &= \frac{\sqrt{-1}}{2} \partial\bar{\partial} \log(1 + |z_1|^2 + \cdots + |z_n|^2) \\
&= \frac{\sqrt{-1}}{2} \partial\bar{\partial} \log\left(1 + \frac{1}{|w_0|^2} + \Sigma \frac{|w_i|^2}{|w_0|^2}\right) \\
&= \frac{\sqrt{-1}}{2} \partial\bar{\partial} \log(1 + |w|^2) - \partial\bar{\partial} \log(|w_0|^2) \\
&= \frac{\sqrt{-1}}{2} \partial\bar{\partial} \log(1 + |w|^2),
\end{aligned}$$

since w_0 is holomorphic. So ω_g is globally defined and the corresponding metric on $\mathbb{C}P^n$ is called the Fubini-Study metric. Clearly, each $\sigma \in SU(n+1)$ acts naturally on $\mathbb{C}P^n$ as an isometry of g, and $SU(n+1)$ acts on $\mathbb{C}P^n$ transitively. Note that

$$\omega_g^n = \left(\frac{\sqrt{-1}}{2}\right)^n \frac{(dz_i \wedge d\bar{z}_i)^n}{(1 + |z|^2)^{n+1}}$$

and hence

$$\mathrm{Ric}_{i\bar{j}} = -\frac{\partial^2}{\partial z_i \partial \bar{z}_j} \log\left(\frac{1}{(1 + |z|^2)^{n+1}}\right) = (n+1) g_{i\bar{j}}.$$

The following calculation

$$\begin{aligned}
-\frac{\partial^2 g_{i\bar{j}}}{\partial z_k \partial \bar{z}_k}|_{z=0} &= -\frac{\partial^4 \log(1+|z|^2)}{\partial z_k \partial \bar{z}_l \partial z_i \partial \bar{z}_j}|_{z=0} \\
&= \frac{1}{2}\frac{\partial^4 |z|^4}{\partial z_k \partial \bar{z}_l \partial z_i \partial \bar{z}_j}|_{z=0} \\
&= \frac{\partial^3}{\partial z_k \partial \bar{z}_l \partial z_i}(|z|^2 z_j)|_{z=0} \\
&= \frac{\partial^2}{\partial z_k \partial z_i}(z_j z_l)|_{z=0} \\
&= \delta_{ij}\delta_{kl} + \delta_{il}\delta_{kj}
\end{aligned}$$

shows us that $\mathbb{C}P^n$ is a manifold of constant bisectional curvature and that $R_{i\bar{j}k\bar{l}} = g_{i\bar{j}}g_{k\bar{l}} + g_{i\bar{l}}g_{k\bar{j}}$, since the isometry group of g acts on $\mathbb{C}P^n$ transitively.

Example 1.10 Let $M = \mathbb{C}^n$ then $\omega_g = \frac{\sqrt{-1}}{2} dz_i \wedge d\bar{z}_i$ is the flat metric and the bisectional curvature vanishes.

Example 1.11 Let $M = B^n = \{z \in \mathbb{C}^n; |z| < 1\}$ and let

$$\omega_g = \frac{\sqrt{-1}}{2}\partial\bar{\partial}\log(1-|z|^2),$$

then $R_{i\bar{j}k\bar{l}} = -(g_{i\bar{j}}g_{k\bar{l}} + g_{i\bar{l}}g_{k\bar{j}})$ and (B^n, g) is Kähler manifold of constant bisectional curvature -1.

Theorem 1.12 *(Uniformization Theorem) If (M, g) is a complete Kähler manifold of constant bisectional curvature $R_{i\bar{j}k\bar{l}} = \lambda(g_{i\bar{j}}g_{k\bar{l}} + g_{i\bar{l}}g_{k\bar{j}})$ for some constant λ, then its universal covering $\widetilde{M}$ is one of the above examples. Moreover, up to scaling, g pulls back to one of the metrics in the above examples.*

Proof. After scaling, we can distinguish three cases, $\lambda = -1$, $\lambda = 0$ and $\lambda = 1$. We will first prove the cases $\lambda \leq 0$. Let (M_λ, g_λ) be the Kähler manifold of constant bisectional curvature λ in the above examples. Consider the maps $\exp_0 : T_0 M_\lambda \to M_\lambda$ and $\exp_p : T_p\widetilde{M} \to \widetilde{M}$, where $\exp_0$ and $\exp_p$ are the exponential maps of g_λ and g, respectively, for any $v \in T_0 M_\lambda$, $\exp_0(v)$ is the geodesic γ of g_λ at time 1, satisfying $\gamma(0) = 0$ and $\gamma'(0) = v$. We need completeness to guarantee the existence of geodesics.

Identify both $T_0 M_\lambda$ and $T_p\widetilde{M}$ with $\mathbb{R}^{2n}$ and define the map

$$\phi = \exp_p(\exp_0)^{-1}.$$

We claim that ϕ is an isometry and once this is proved, we have proved the theorem. So we must consider $d\phi$. We will first look at $d\exp_0(v)$. For $w \in \mathbb{R}^{2n}$, we have

$$d\exp_0(v)(w) = \frac{\partial}{\partial t}\exp_0(s(v+tw)) = \frac{\partial}{\partial t}\gamma(s,t) =: X_w(s)$$

where we have introduced $\gamma(s,t) = \exp_0(s(v+tw))$ for simplicity.

We now check that $X_w(s)$ is a Jacobi field. Because $\gamma(s,t)$ is a geodesic for each fixed t, we have that $\nabla_{\frac{\partial}{\partial s}\gamma(s,t)}\frac{\partial}{\partial s}\gamma(s,t) = 0$ and hence $\nabla_{\frac{\partial}{\partial t}}\nabla_{\frac{\partial}{\partial s}}\frac{\partial}{\partial s}\gamma(s,t) = 0$ and then

$$\begin{aligned} 0 &= \nabla_{X_w}\nabla_{\frac{\partial\gamma}{\partial s}}\frac{\partial\gamma}{\partial s} \\ &= \nabla_{\frac{\partial\gamma}{\partial s}}\nabla_{X_w}\frac{\partial\gamma}{\partial s} + \nabla_{[X_w,\frac{\partial\gamma}{\partial s}]}\frac{\partial\gamma}{\partial s} - R\left(\frac{\partial\gamma}{\partial s}, X_w\right)\frac{\partial\gamma}{\partial s} \\ &= \nabla_{\frac{\partial\gamma}{\partial s}}\nabla_{\frac{\partial\gamma}{\partial s}}X_w - R\left(\frac{\partial\gamma}{\partial s}, X_w\right)\frac{\partial\gamma}{\partial s} \end{aligned}$$

together with the initial conditions $X_w(0) = 0$ and $X'_w(0) = w$. Now fix an orthonormal basis $e_1, \ldots, e_{2n}$ for T_0M_λ, such that $e_1 = \frac{\partial\gamma}{\partial t}|\frac{\partial\gamma}{\partial t}|^{-1}$ and $e_{n+i} = Je_i$. Parallel transport this basis along γ, that is, $\nabla_{\frac{\partial\gamma}{\partial t}}e_i(s) = 0$ and $e_i(0) = e_i$. So we can write $X_w(s) = \sum X^i(s)e_i(s)$ and the Jacobi equation reads

$$\frac{\partial^2 X^i(s)}{\partial^2 s}e_i - R\left(\frac{\partial\gamma}{\partial s}, e_j\right)\frac{\partial\gamma}{\partial s}X^j = 0.$$

Taking the inner product with e_i, we get

$$\frac{\partial^2 X^i(s)}{\partial^2 s} - \left\langle R\left(\frac{\partial\gamma}{\partial s}, e_j\right)\frac{\partial\gamma}{\partial s}, e_i\right\rangle X^j = 0$$

which is equivalent to

$$\frac{\partial^2 X^i(s)}{\partial^2 s} + \left|\frac{\partial\gamma}{\partial s}\right|^2 R(e_1, e_j, e_i, e_1)X^j = 0.$$

By the lemma below, the $R(e_1, e_j, e_i, e_1)$ are completely determined by the bisectional curvature, so X_w is uniquely determined by w and λ. If now $\widetilde{X}_w(s) = d\exp_p(v)(w)$ is a Jacobi field along the corresponding geodesic $\tilde{\gamma}$ in $\widetilde{M}$ then $\widetilde{X}_w$ satisfies the same equation as X_w does because the bisectional curvatures are the same and hence we can deduce that $|X_w| = |\widetilde{X}_w|$, so ϕ is an isometry.

If $\lambda > 0$ we know that the Ricci curvature is also positive and by Myers' Theorem [5], we know that $\widetilde{M}$ is compact. Let $U_0 \subset \mathbb{C}P^n$ as before then we can show that ϕ is an isometry from U_0 onto its image in exactly the same way as before. Now because U_0 is dense in $\mathbb{C}P^n$ and because $\widetilde{M}$ is compact we can extend ϕ to all of $\mathbb{C}P^n$ so that ϕ remains an isometry. □

Lemma 1.13 *If the bisectional curvature is constant, then*

$$R(e_1, e_i)e_1 = \begin{cases} -\frac{\lambda}{2}e_i, & \text{if } e_i \perp e_1, Je_1 \\ -2\lambda e_i, & \text{if } e_i = Je_1. \end{cases}$$

Proof. As before, we write $u_i = \frac{1}{\sqrt{2}}(e_i - \sqrt{-1}Je_i)$. Then

$$R(u_i, \bar{u}_i, u_k, \bar{u}_l) = -R(e_i, Je_i, e_k, Je_l) + \sqrt{-1}R(e_i, Je_i, e_k, e_l).$$

Since the bisectional curvature is constant λ, we have

$$-R(e_i, Je_i, e_k, Je_l) + \sqrt{-1}R(e_i, Je_i, e_k, e_l) = \begin{cases} 2\lambda\delta_{kl}, & \text{if } k = i \\ \lambda\delta_{kl}, & \text{if } k \neq i. \end{cases}$$

Comparing the real and imaginary parts of both sides, we get

$$-R(e_i, Je_i, e_k, Je_l) = \begin{cases} 2\lambda\delta_{kl}, & \text{if } k = i \\ \lambda\delta_{kl}, & \text{if } k \neq i. \end{cases}$$

and

$$R(e_i, Je_i, e_k, e_l) = 0.$$

For any $i \neq j$, if we replace e_i by $\frac{1}{\sqrt{2}}(e_i + e_j)$ in the above arguments, we get

$$R(e_i, Je_j, e_k, Je_l) = R(e_i, Je_j, e_k, e_l) = 0.$$

It follows from the above that

$$R(e_1, e_i, e_1, Je_j) = 0, \quad R(e_1, e_i, e_1, e_l) = 0, \, l \neq i.$$

So $R(e_1, e_i)e_1 = \mu e_i$ for some μ. If $e_i = Je_1$, then $\mu = -2\lambda$, since

$$R(e_1, Je_1, Je_1, e_1) = R(u_1, \bar{u}_1, u_1, \bar{u}_1) = 2\lambda.$$

Now we may assume that $e_i \perp e_1, Je_1$. Replace e_1, Je_1 by $\cos\theta e_1 + \sin\theta Je_1$, $-\sin\theta e_1 + \cos\theta Je_1$, as above, we can deduce

$$R(\cos\theta e_1 + \sin\theta Je_1, e_i, J(\cos\theta e_1 + \sin\theta Je_1), e_i) = 0.$$

This is the same as

$$\frac{d}{d\theta}R(\cos\theta e_1 + \sin\theta Je_1, e_i, J(\cos\theta e_1 + \sin\theta Je_1), e_i) = 0.$$

Hence, $R(e_1, e_i, e_1, e_i) = R(Je_1, e_i, Je_1, e_i)$, then by $R(u_1, \bar{u}_1, u_1, \bar{u}_1) = \lambda$, we have $\mu = -\frac{\lambda}{2}$. Then the lemma is proved. □

Chapter 2

Extremal Kähler metrics

2.1 The space of Kähler metrics

In this section, we introduce the Calabi functional on the space of Kähler metrics. We will start with a simple lemma.

Lemma 2.1 *($\partial\bar{\partial}$-Lemma) Let (M,g) be a Kähler manifold and let $\phi_1, \phi_2 \in H^{1,1}(M,\mathbb{C})$ and suppose that ϕ_1 is cohomologous to ϕ_2. Then there exists a function $f \in C^\infty(M,\mathbb{R})$ such that $\phi_1 - \phi_2 = \partial\bar{\partial} f$.*

Proof. We know that $\phi_1 - \phi_2 = d\psi$ for some $\psi \in H^1(M,\mathbb{R})$ and that we can write $\psi = \psi^{1,0} + \psi^{0,1}$ with $\bar{\partial}\psi^{0,1} = \partial\psi^{1,0} = 0$, so we want to find functions $f, \tilde{f}$ satisfying $\psi^{0,1} = \bar{\partial} f$ and $\psi^{1,0} = \partial \tilde{f}$. We will show how to do this for $\psi^{0,1}$ and by taking conjugates, one can do this also for $\psi^{1,0}$. Write $\psi^{0,1} = \theta$, then locally,

$$\theta = \theta_{\bar{j}} d\bar{z}_j \quad \text{and} \quad \bar{\partial}^* \theta = -g^{i\bar{j}} \theta_{\bar{j}i}.$$

The following equation can be solved for u,

$$\bar{\partial}^* \theta = \bar{\partial}^* \bar{\partial} u = -g^{i\bar{j}} \frac{\partial^2 u}{\partial z_i \partial z_{\bar{j}}},$$

because $\int_M \bar{\partial}^* \theta \omega_g^n = 0$. So $\bar{\partial}^*(\theta - \bar{\partial} u) = 0$ and we also have that $\bar{\partial}(\theta - \bar{\partial} u) = 0$. We claim that $\partial(\theta - \bar{\partial} u) = 0$. Suppose this claim were true. Then writing $\psi^{0,1} = \theta - \bar{\partial} u + \bar{\partial} u$, we see that $d\psi^{0,1} = \partial\bar{\partial} u$, which proves the lemma. So it remains to prove the claim. Set $\phi = \theta - \bar{\partial} u$, then we have the identity

$$\left(\frac{\sqrt{-1}}{2}\right)^2 \partial\phi \wedge \overline{\partial\phi} \wedge \omega_g^{n-2} = \frac{1}{n(n-1)} \left(|\partial\phi|^2 - |\bar{\partial}^*\phi|^2 \right) \omega_g^n.$$

Integrating this, we get

$$\begin{aligned}
0 &\le \int_M |\partial\phi|^2 \omega_g^n \\
&= \int_M |\bar\partial^*\phi|^2\omega_g^n - \frac{n(n-1)}{4}\int_M \partial\phi\wedge\overline{\partial\phi}\wedge\omega_g^{n-2} \\
&= \int_M |\bar\partial^*\phi|^2\omega_g^n - \frac{n(n-1)}{4}\int_M \bar\partial\left(\partial\phi\wedge\overline{\phi}\wedge\omega_g^{n-2}\right) \\
&\quad - \frac{n(n-1)}{4}\int_M \partial\bar\partial\phi\wedge\overline{\phi}\wedge\omega_g^{n-2}.
\end{aligned}$$

Therefore, $\partial\phi = 0$, so the lemma is proved. □

Note that this lemma is also true for any (p,q)-forms (compare [3]).

Corollary 2.2 *Given* $\Omega \in H^{1,1}(M,\mathbb{C}) \cap H^2(M,\mathbb{R})$, *define*

$$\mathcal{K}_\Omega = \{\text{all Kähler metrics } \omega \text{ with } [\omega] = \Omega\},$$

then

$$\begin{aligned}
\mathcal{K}_\Omega &= \{\omega_g + \frac{\sqrt{-1}}{2}\partial\bar\partial\phi;\ \phi\in C^\infty(M,\mathbb{R}), \int_M \phi\,\omega_g^n = 0\} \\
&\simeq \{\phi \in C^\infty(M,\mathbb{R}); \int_M \phi\,\omega_g^n = 0,\ \omega_g + \frac{\sqrt{-1}}{2}\partial\bar\partial\phi > 0\}.
\end{aligned}$$

In what follows we will use results and definitions from Calabi (1980) in [4].

Definition 2.3 *The scalar curvature of a metric* ω *is defined to be*

$$s(\omega) = R_{i\bar{i}j\bar{j}} = -g^{i\bar{j}}\frac{\partial^2}{\partial z_i\partial\bar{z}_j}\log\det g_{k\bar{l}}.$$

Definition 2.4 *The Calabi functional* Ca *is given by*

$$Ca(\omega) = \frac{1}{V}\int_M s(\omega)^2\omega^n$$

where $V = \int_M \omega^n$.

Note that V only depends on the class of ω and is independent of the particular representative that we choose. Fix now a metric $\omega = \frac{\sqrt{-1}}{2} g_{i\bar{j}} dz_i \wedge d\bar{z}_j$

and calculate the Euler-Lagrange equations for the Calabi functional. Let $\omega_{t\phi} = \omega + \frac{t\sqrt{-1}}{2}\partial\bar\partial\phi$, then

$$\begin{aligned} s(\omega_{t\phi}) &= -g_{t\phi}^{i\bar j}\frac{\partial^2}{\partial z_i\partial\bar z_j}\log\det\left(g_{k\bar l} + t\frac{\partial^2\phi}{\partial z_k\partial\bar z_l}\right) \\ &= s(\omega) + t\left(-g^{i\bar j}\frac{\partial^2}{\partial z_i\partial\bar z_j}\Delta\phi - g^{i\bar l}\frac{\partial^2\phi}{\partial z_k\partial\bar z_l}g^{k\bar j}R_{i\bar j}\right) + \mathcal{O}(t^2). \end{aligned}$$

And so we get

$$Ca(\omega_{t\phi}) = Ca(\omega) + \frac{t}{V}\int_M 2s(\omega)\left(-\Delta^2\phi - \phi_{k\bar l}R_{l\bar k} + \frac{s(\omega)}{2}\Delta\phi\right)\omega^n + \mathcal{O}(t^2).$$

Here the last term under the integral comes from the fact that we have also changed the volume form and we have

$$\frac{d}{dt}\left(\omega + \frac{t\sqrt{-1}}{2}\partial\bar\partial\phi\right)^n|_{t=0} = n\,\frac{\sqrt{-1}}{2}\,\omega^{n-1}\wedge\partial\bar\partial\phi = \Delta\phi\,\omega^n.$$

Integration by parts gives us then the following Euler-Lagrange equation for the Calabi functional.

$$-2\Delta^2 s - 2(sR_{l\bar k})_{k\bar l} + \Delta(s^2) = 0.$$

Here, for any covariant tensor ϕ given by $\{\phi_{i_1\ldots i_p\bar j_1\ldots\bar j_q}\}$ in local coordinates, we define

$$\phi_{i_1\ldots i_p\bar j_1\ldots\bar j_q,i} = \frac{\partial\phi_{i_l\ldots i_p\bar j_l\ldots\bar j_q}}{\partial z_i} - \sum_{\alpha=1}^{k}\Gamma^j_{ii_\alpha}\phi_{i_1\ldots\underset{(\alpha)}{i}\ldots i_p\bar j_1\ldots\bar j_q},$$

these $\phi_{i_1\ldots i_p\bar j_1\ldots\bar j_q,i}$ are components of the covariant derivative $\nabla^{1,0}\phi$ of ϕ. Similarly, one can define $\phi_{i_1\ldots i_p\bar j_1\ldots\bar j_q,\bar j}$ representing $\nabla^{0,1}\phi$. We know

$$\Delta(s^2) = 2|\nabla s|^2 + 2s\Delta s.$$

So we can use the second Bianchi identity

$$R_{ijkl,m} + R_{ijlm,k} + R_{ijmk,l} = 0$$

to reduce the above to

$$\Delta^2 s + |\nabla s|^2 + R_{i\bar j}s_{\bar i j} = 0.$$

This is still an equation of too high order, but now using that M is compact, we can reduce the equation to a second order one in s. We will show that the above equation is equivalent to $s_{\bar{i}\bar{j}ji} = 0$, which again follows from taking and switching some derivatives and using the second Bianchi identity, namely

$$s_{\bar{i}\bar{j}ji} = s_{\bar{j}\bar{i}ji} = s_{\bar{j}j\bar{i}i} + (R_{k\bar{j}\bar{i}j}s_{\bar{k}})_i = 0.$$

And so we see that

$$0 = \int_M s_{\bar{i}\bar{j}ji} s\omega^n = \int_M s_{\bar{i}\bar{j}} s_{ij}\omega^n = \int_M |s_{\bar{i}\bar{j}}|^2\omega^n.$$

Therefore, $s_{\bar{i}\bar{j}} = 0$.

Definition 2.5 *A Kähler metric ω is called extremal if $s(\omega)_{\bar{i}\bar{j}} = 0$.*

We can also give this a geometrical interpretation: Define a vector field on M by $X^i = g^{i\bar{j}}s_{\bar{j}}$. Then ω is extremal is equivalent to saying that $\bar{\partial}X = 0$ (since the metric is parallel) and this is equivalent to saying that X is a holomorphic vector field.

Corollary 2.6 *If M has no non-zero holomorphic vector fields, then the extremal metrics are the metrics with constant scalar curvature.*

2.2 A brief review of Chern classes

Given a Kähler metric ω_g, we can define a matrix valued 2-form $\Omega = (\Omega_i^j)$, which is actually of type (1,1), by

$$\Omega_i^j = g^{j\bar{p}}R_{i\bar{p}k\bar{l}}dz_k \wedge d\bar{z}_l.$$

Then we have that $\det(I + \frac{t\sqrt{-1}}{2\pi}\Omega) = 1 + t\phi_1(g) + t^2\phi_2(g) + \cdots$ which has the following well-known properties (compare [3]):

- $d\phi_i(g) = 0$ and $[\phi_i] \in H^{i,i}(M,\mathbb{C}) \cap H^{2i}(M,\mathbb{R})$,
- $[\phi_i(g)]$ is independent of g,
- $c_i(M)_{\mathbb{R}}$ is represented by $\phi_i(g)$.

Here the $c_i(M)$ denotes the ith Chern class of the manifold M. We will mainly be interested in $c_1(M)$ and $c_2(M)$ and will only check the second property for $c_1(M)$.

$$\phi_1(g) = \frac{\sqrt{-1}}{2\pi}\Omega_i^i = \frac{\sqrt{-1}}{2\pi}R_{k\bar{l}}dz_k \wedge d\bar{z}_l = -\frac{\sqrt{-1}}{2\pi}\partial\bar{\partial}\log\det(g_{k\bar{l}})$$

and hence

$$\phi_1(g) - \phi_1(g') = -\frac{\sqrt{-1}}{2\pi}\partial\bar{\partial}\log\left(\frac{\det(g_{k\bar{l}})}{\det(g'_{k\bar{l}})}\right).$$

Clearly, $\log\left(\frac{\det(g_{k\bar{l}})}{\det(g'_{k\bar{l}})}\right)$ is a well-defined function. Moreover, if

$$dV = v_z dz_1 \wedge d\bar{z}_1 \wedge \cdots dz_n \wedge d\bar{z}_n$$

is any volume form on M, we can also represent $c_1(M)$ by

$$-\partial\bar{\partial}\log v_z.$$

Definition 2.7 *We say that $c_1(M) > 0$ (< 0) if $c_1(M)$ can be represented by a positive (negative) form. In local coordinates, this means that*

$$\phi = \sqrt{-1}\phi_{k\bar{l}}dz_k \wedge d\bar{z}_l$$

where $\phi_{k\bar{l}}$ is positive (negative) definite. We say that $c_1(M) = 0$ if the first Chern class $c_1(M)$ is cohomologous to zero.

Example 2.8 Let $M = \mathbb{C}P^n$ and let

$$\omega_{FS} = \frac{\sqrt{-1}}{2}\partial\bar{\partial}\log(1 + \sum |z_i|^2) > 0,$$

then we have

$$\omega_{FS}^n = \left(\frac{\sqrt{-1}}{2}\right)^n \frac{n! dz_1 \wedge \cdots \wedge d\bar{z}_n}{(1 + \sum |z_i|^2)^{n+1}}.$$

This implies that

$$\mathrm{Ric}(\omega_{FS}) = \frac{n+1}{\pi}\omega_{FS} \quad \text{and so} \quad c_1(M) = \frac{n+1}{\pi}[\omega_{FS}],$$

where $\frac{1}{\pi}[\omega_{FS}]$ is the positive generator of the cohomology $H^2(\mathbb{C}P^n, \mathbb{Z}) = \mathbb{Z}$. One can show that

$$\frac{2(n+1)}{n}c_2(M) = c_1(M)^2.$$

Here we mean by squaring that we take the cup product of $c_1(M)$ with itself. This implies that

$$\int_M \left(\frac{2(n+1)}{n}c_2(M) - c_1(M)^2\right) \wedge \omega_{FS}^{n-2} = 0.$$

As a side remark, we would like to mention here the concept of an index of a symplectic manifold (M, ω). Recall that a symplectic manifold (M, ω) is a differentiable manifold M with a non-degenerate, d-closed 2-form ω. The index of M is r if $c_1(M) = r\phi$ for some primitive symplectic form ϕ, that is, ϕ is not an integer multiple of another form. So the index of $\mathbb{C}P^n$ is $n + 1$.

For $n \geq 1$, one can prove that for an n-dimensional Kähler manifold M, $r \leq n+1$, and the equality holds if and only if $M = \mathbb{C}P^n$ and that $r = n$ if and only if M is a quadratic hypersurface in $\mathbb{C}P^n$. For $n = 2$, this is a classical result, the case $n > 2$ was proved by S. Kobayashi and Ochiai (see [15]). A plausible conjecture is that *the index of any compact symplectic manifold of complex dimension n is no more than $n + 1$.* It may also be true that if a compact symplectic manifold of complex dimension n is of index $n+1$, then it is $\mathbb{C}P^n$. This conjecture is indeed true in the case that $n = 2$ and the underlying manifold is already diffeomorphic to $\mathbb{C}P^2$. This was a result of C. Taubes.

Example 2.9 Let $M_f = \{z \in \mathbb{C}P^{n+1}; f(z) = 0\}$ where f is a homogeneous polynomial of degree d. It follows from the Implicit Function Theorem that M_f is a smooth manifold if and only if

$$\cap_{i=0}^{n+1}\left\{\frac{\partial f}{\partial z_i} = 0\} = \{(0,\ldots,0)\right\}.$$

So assume this is the case. We have seen before that we can equip $\mathbb{C}P^{n+1}$ with the Fubini-Study metric so that

$$g_{i\bar{j}} = \frac{\partial^2}{\partial z_i \partial \bar{z}_j}\log(1+|z|^2).$$

We claim that

$$c_1(M_f) = \frac{n+2-d}{\pi}[\omega_{FS}].$$

Set

$$\psi = \log\left(\frac{\sum|\frac{\partial f}{\partial z_i}|^2}{|z|^{2(d-1)}}\right)$$

and note that ψ is globally defined. We will now calculate ω_{FS}^n. It suffices to do this on the open and dense subset

$$U_0 \cap \left\{\frac{\partial f}{\partial z_{n+1}} \neq 0\right\} \cap M_f,$$

so we get

$$\left(\frac{2}{\sqrt{-1}}\right)^n \frac{\omega_{FS}^n}{n!}$$
$$= \sum_{i,j=1}^{n+1}(-1)^{i+j}\det(g_{\alpha\bar{\beta}})_{1\leq\alpha,\beta\leq n+1,\alpha\neq i,\beta\neq j}$$
$$dz_1 \wedge d\bar{z}_1 \wedge \cdots \wedge \widehat{dz_i} \wedge d\bar{z}_i \wedge \cdots \wedge dz_j \wedge \widehat{d\bar{z}_j} \wedge \cdots \wedge dz_{n+1} \wedge d\bar{z}_{n+1}$$

$$
\begin{aligned}
&= \det(g_{\alpha\bar\beta})_{1\le\alpha,\beta\le n} dz_1\wedge d\bar z_1\wedge\cdots\wedge dz_n\wedge d\bar z_n\\
&\quad + 2\mathrm{Re}(\sum_{i=1}^{n}(-1)^{n+i+1}\det(g_{\alpha\bar\beta})_{1\le\alpha\le n+1,\alpha\ne i,1\le\beta\le n}\\
&\quad dz_1\wedge d\bar z_1\wedge\cdots\wedge \widehat{dz_i}\wedge d\bar z_i\wedge\cdots\wedge dz_n\wedge d\bar z_n\wedge dz_{n+1}\\
&\quad + \sum_{i,j=1}^{n}(-1)^{i+j}\det(g_{\alpha\bar\beta})_{1\le\alpha,\beta\le n,\alpha\ne i,\beta\ne j}\\
&\quad dz_1\wedge d\bar z_1\wedge\cdots\wedge \widehat{dz_i}\wedge\cdots\wedge dz_j\wedge \widehat{d\bar z_j}\wedge\cdots\wedge dz_{n+1}\wedge d\bar z_{n+1}\\
&= \left(\det(g_{\alpha\bar\beta})_{1\le\alpha,\beta\le n} - 2\mathrm{Re}\left(\sum_{i=1}^{n}(-1)^{n+i+1}\det(g_{\alpha\bar\beta})_{1\le\alpha\le n+1,\alpha\ne i,1\le\beta\le n}\frac{\partial z_{n+1}}{\partial z_j}\right)\right.\\
&\quad \left. + \sum_{i,j=1}^{n}(-1)^{i+j}\det(g_{\alpha\bar\beta})_{1\le\alpha,\beta\le n,\alpha\ne i,\beta\ne j}\overline{\frac{\partial z_{n+1}}{\partial z_j}}\frac{\partial z_{n+1}}{\partial z_i}\right)\\
&\quad dz_1\wedge d\bar z_1\wedge\cdots\wedge dz_n\wedge d\bar z_n.
\end{aligned}
$$

Using now that $\frac{\partial z_{n+1}}{\partial z_i} = -\frac{\partial f}{\partial z_i}(\frac{\partial f}{\partial z_{n+1}})^{-1}$, the above expression reduces to

$$
\left(\frac{2}{\sqrt{-1}}\right)^n \frac{\omega_{FS}^n}{n!} = |\frac{\partial f}{\partial z_{n+1}}|^{-2}\det(g_{i\bar j})_{1\le i,j\le n+1}\sum_{i,j=1}^{n+1} g^{i\bar j}\frac{\partial f}{\partial z_i}\overline{\frac{\partial f}{\partial z_j}} dz_1\wedge\cdots d\bar z_n.
$$

Now recall that

$$
\begin{aligned}
g_{i\bar j} &= \frac{1}{1+|z|^2}\left(\delta_{ij} - \frac{z_i\bar z_j}{1+|z|^2}\right)\\
g^{i\bar j} &= (1+|z|^2)(\delta_{ij} + z_i\bar z_j),
\end{aligned}
$$

so it follows that

$$
\sum_{i,j=1}^{n} g^{i\bar j}\frac{\partial f}{\partial z_i}\overline{\frac{\partial f}{\partial z_j}} = (1+|z|^2)\left(\sum_{i=1}^{n+1}|\frac{\partial f}{\partial z_i}|^2 + \sum_{i,j=1}^{n+1} z_i\frac{\partial f}{\partial z_i}\overline{z_j\frac{\partial f}{\partial z_j}}\right).
$$

But now we can use that on M_f,

$$
0 = df = \frac{\partial f}{\partial z_0} + \sum_{i=1}^{n+1} z_i\frac{\partial f}{\partial z_i},
$$

and therefore

$$
\sum_{i,j=1}^{n} g_{i\bar j}\frac{\partial f}{\partial z_i}\overline{\frac{\partial f}{\partial z_j}} = (1+|z|^2)\left(\sum_{i=0}^{n+1}|\frac{\partial f}{\partial z_i}|^2\right).
$$

Because the determinant of the metric is given by

$$\det g_{i\bar{j}} = \frac{1}{(1+|z|^2)^{n+2}},$$

we see that

$$\begin{aligned} -\partial\bar{\partial}\log\omega_{FS}^n &= -\partial\bar{\partial}\log\frac{\sum|\frac{\partial f}{\partial z_i}|^2}{(1+|z|^2)^{n+1}} - \partial\bar{\partial}\log|\frac{\partial f}{\partial z_{n+1}}|^2 \\ &= -\partial\bar{\partial}\log e^{\psi}(1+|z|^2)^{d-n-2} \\ &= -\partial\bar{\partial}\psi + \frac{2(n-d+2)}{\sqrt{-1}}\omega_{FS}. \end{aligned}$$

Because the above expression equals the Ricci curvature, we have proved the above claim about the first Chern class.

It follows that the first Chern class $c_1(M_f)$ is positive, zero or negative according to whether $d < n+2$, $d = n+2$ or $d > n+2$.

2.3 Uniformization of Kähler-Einstein manifolds

In this section, we collect a few facts on Kähler-Einstein manifolds (compare [3], [27]). The main theorem gives a characterization of Kähler manifolds of constant curvature in terms of Chern numbers.

First we introduce

Definition 2.10 *We say that g is a Kähler-Einstein metric if there exists a real constant λ such that* $\mathrm{Ric}(g) = \lambda\omega_g$, *where* $\mathrm{Ric}(g)$ *is the Ricci form defined in any local coordinates as*

$$\frac{\sqrt{-1}}{2}\sum_{i,j} R_{i\bar{j}}dz_i \wedge d\bar{z}_j.$$

A Kähler manifold (M, g) is Kähler-Einstein if g is a Kähler-Einstein metric.

For simplicity, we will denote a Kähler metric by its Kähler form.

Lemma 2.11 *The average of the scalar curvature depends only on $[\omega_g]$ and $c_1(M)$.*

Proof. Given any point $x \in M$, we can choose coordinates $z_1, \ldots, z_n$ such that at x, we have

$$\omega_g = \frac{\sqrt{-1}}{2}\sum dz_i \wedge d\bar{z}_i$$

and

$$\mathrm{Ric}(g) = \frac{\sqrt{-1}}{2}\sum R_{i\bar{i}}dz_i \wedge d\bar{z}_i.$$

It follows that at x,

$$\begin{aligned}&\operatorname{Ric}(g)\wedge\omega_g^{n-1}\\&=(n-1)!\left(\frac{\sqrt{-1}}{2}\right)^n\sum_i R_{i\bar{i}}\,dz_1\wedge d\bar{z}_1\wedge\cdots\wedge dz_n\wedge d\bar{z}_n\\&=\frac{1}{n}s(\omega_g)\omega_g^n.\end{aligned}$$

So we get

$$\pi c_1(M)[\omega_g]^{n-1}=\frac{1}{n}\int_M s(\omega_g)\omega_g^n.$$

□

Proposition 2.12 *If $\pi c_1(M)=\lambda[\omega]$ and ω is a Kähler metric with constant scalar curvature, then ω is Kähler-Einstein.*

Proof. We know that $\frac{1}{\pi}\operatorname{Ric}(\omega)$ represents the first Chern class of M and that it is a $(1,1)$-form. So by the $\partial\bar{\partial}$-Lemma, we have that $\operatorname{Ric}(\omega)-\lambda\omega=\partial\bar{\partial}f$. Now taking the trace of this, we have that $s(\omega)-n\lambda=\Delta f$, but note now that by the above lemma, $s(\omega)=n\lambda$, so f is a harmonic function on a compact manifold and therefore f is constant. □

Theorem 2.13 *Given any Kähler-Einstein manifold (M,g). Then $(\widetilde{M},\omega)\simeq\mathbb{C}P^n,\mathbb{C}^n$ or B^n if and only if*

$$\left(\frac{2(n+1)}{n}c_2(M)-c_1(M)^2\right)[\omega]^{n-2}=0.$$

Here by $\simeq$, we mean isometric up to scaling.

Proof. By Theorem 1.10, it suffices to prove that the last equality is equivalent to M being a manifold of constant bisectional curvature, that is, there exists a constant λ such that $R_{i\bar{j}k\bar{l}}=\lambda(g_{i\bar{j}}g_{k\bar{l}}+g_{i\bar{l}}g_{k\bar{j}})$. Recall that we defined the Chern classes by taking

$$\det\left(I+\frac{t\sqrt{-1}}{2\pi}\Omega\right)=I+t\phi_1(\Omega)+t^2\phi_2(\Omega)+\cdots,$$

where Ω is defined in last section. The ϕ_i represent the Chern classes of M, in particular, $\phi_1(\Omega)=\operatorname{tr}\Omega$ represents the first Chern class.

Viewing Ω as a matrix valued 2-form and using the properties of trace and determinant, we see that

$$\phi_1(\Omega)^2-2\phi_2(\Omega)=\left(\frac{\sqrt{-1}}{2\pi}\right)^2\operatorname{tr}(\Omega\wedge\Omega).$$

So we have that

$$(c_1(M)^2 - 2c_2(M))[\omega]^{n-2} = \left(\frac{\sqrt{-1}}{2\pi}\right)^2 \int_M \operatorname{tr}(\Omega \wedge \Omega) \wedge \omega^{n-2}.$$

Denote by $R^o(g)$ the traceless part of the curvature $R(g)$. Because M is Kähler-Einstein, we have $\operatorname{Ric}(g) = \lambda\omega_g$ and so in local coordinates, we get

$$R^o_{i\bar{j}k\bar{l}} = R_{i\bar{j}k\bar{l}} - \frac{\lambda}{n+1}(g_{i\bar{j}}g_{k\bar{l}} + g_{i\bar{l}}g_{k\bar{j}}).$$

The tensor R^o measures how much the metric deviates from having constant bisectional curvature. So it now suffices to show that $R^o = 0$ is equivalent to the equality in the statement of the theorem. If we denote by $|-|$ the norm given by the metric, then by direct computations, we have

$$\frac{1}{n(n-1)}|R^o|^2\omega_g^n = -\left(\frac{\sqrt{-1}}{2}\right)^2 \operatorname{tr}(\Omega \wedge \Omega) \wedge \omega_g^{n-2} + \frac{\lambda^2}{n+1}\omega_g^n.$$

It follows that the Chern number

$$\pi^2(c_1(M)^2 - 2c_2(M))[\omega_g]^{n-2} = \int_M \frac{\lambda^2}{n+1}\omega_g^n - \frac{1}{n(n-1)}\int_M |R^o|^2\omega_g^n.$$

We also know that

$$\int_M \lambda^2\omega_g^n = \pi^2 c_1(M)^2[\omega_g]^{n-2}$$

and therefore

$$\frac{n\pi^2}{n+1}\left(c_1(M)^2 - \frac{2(n+1)}{n}c_2(M)\right)[\omega_g]^{n-2} = -\frac{1}{n(n-1)}\int_M |R^o|^2\omega_g^n.$$

So we have shown that M is a manifold of constant bisectional curvature if and only if

$$(c_1(M)^2 - \frac{2(n+1)}{n}c_2(M))[\omega]^{n-2} = 0.$$

More generally, we have shown that for any Kähler-Einstein manifolds, not necessarily of constant bisectional curvature, the following Chern number inequality holds (cf. [28]).

$$\left(c_1(M)^2 - \frac{2(n+1)}{n}c_2(M)\right)[\omega]^{n-2} \leq 0.$$

□

Note that if $\lambda = \pm 1$, we have that $[\omega_g] = \pm\pi c_1(M)$, but if $\lambda = 0$, we have more freedom in choosing $[\omega_g]$.

Remark 2.14 Any Kähler manifold M with $c_1(M) > 0$ is simply connected. It follows from Yau's solution of the Calabi conjecture (see Chapter 5) that there is always a Kähler metric ω with positive Ricci curvature on M. It follows then from Kobayashi (see [13]) that M is simply connected. This is because $\mathrm{Ric}(\omega) > 0$ implies that the fundamental group of M is finite. But for a compact Kähler manifold with positive first Chern class, we know that $h^{q,0} = 0$ for $q = 1, \dots, n$. Defining the holomorphic Euler characteristic by $\chi(M) = \sum_{i=0}^{n}(-1)^q h^{q,0}$, then a compact Kähler manifold with $c_1(M) > 0$ has holomorphic Euler characteristic equal to 1. If $\tilde{M}$ is the k-fold covering of M, then $\chi(\tilde{M}) = k\chi(M)$, but because both M and $\tilde{M}$ have positive first Chern class, they have the same holomorphic Euler characteristic and therefore we have a one-fold covering. This then implies that the fundamental group of M has no proper subgroups of finite index. Since the group is finite, this implies that M is simply connected.

Remark 2.15 In general, if (M, g) is a compact Kähler manifold with the Kähler form ω_g and constant scalar curvature $s(\omega_g)$, then

$$2\pi^2\Big(c_1(M)^2 - c_2(M)\Big) \cdot [\omega_g]^{n-2} \leq \frac{n+2}{n^2(n+1)} s(\omega_g)^2[\omega_g]^n,$$

and the equality holds if and only if g is of constant bisectional curvature.

Chapter 3

Calabi-Futaki invariants

3.1 Definition of Calabi-Futaki invariants

In this section, we will introduce a holomorphic invariant for Kähler manifolds, which was first done by Futaki for manifolds with positive first Chern class and then by Calabi and Futaki for general Kähler manifolds. We will take a slightly different approach from the original one taken.

Definition 3.1 *We define the Kähler cone $Ka(M)$ to be the set of all cohomology classes $[\omega] \in H^2(M,\mathbb{R}) \cap H^{1,1}(M,\mathbb{C})$ which can be represented by a Kähler metric.*

Let $\eta(M)$ be the space of all holomorphic vector fields on M. That is, all sections $X \in \Gamma(T_{\mathbb{C}}M)$, which can locally be written as $X = X^i \frac{\partial}{\partial z_i}$, where the X^i are holomorphic functions on M. We will define a complex valued function f_M on $Ka(M) \times \eta(M)$ in the following way: Given any $[\omega] \in Ka(M)$ and any $X \in \eta(M)$, pick a Kähler metric $\omega_g \in [\omega]$. Let h_g be the function defined by

$$s(g) - \frac{1}{V}\int_M s(g)\omega_g^n = \Delta_g h_g.$$

This is well defined since integrating the left-hand side of this expression over M equals 0 and h_g is unique up to addition of a constant. Using results from the last section, we can define

$$\mu := \frac{1}{V}\int_M s(g)\omega_g^n = \frac{n\pi c_1(M)[\omega]^{n-1}}{[\omega]^n}.$$

Define the Calabi-Futaki invariant f_M by

$$f_M([\omega], X) = \int_M X(h_g)\omega_g^n.$$

Next we will prove that the Calabi-Futaki invariant is well defined and does not depend on the particular representative we choose. We will need various steps to prove this. First note that because ω_g is a closed form and because X is holomorphic, we have that $\bar{\partial}(i_X\omega_g) = 0$. It follows from the Hodge Theorem that we can find a smooth function θ_X and a harmonic 1-form α such that

$$i_X\omega_g = \alpha - \bar{\partial}\theta_X.$$

Clearly, $\bar{\partial}\alpha = 0$ and we also have that $\bar{\partial}^*\alpha = 0$, so we see that

$$\begin{aligned} f_M([\omega], X) &= \int_M X(h_g)\omega_g^n \\ &= \int_M X^i \frac{\partial h_g}{\partial z_i}\omega_g^n \\ &= \int_M \left(g^{i\bar{j}}\alpha_{\bar{j}} \frac{\partial h_g}{\partial z_i} - g^{i\bar{j}} \frac{\partial \theta_X}{\partial \bar{z}_j} \frac{\partial h_g}{\partial z_i} \right) \omega_g^n \\ &= \int_M h_g \Delta_g \theta_X \omega_g^n, \end{aligned}$$

and therefore, without loss of generality, we may assume that $\alpha = 0$. So from now on, we have that $i_X\omega_g = -\bar{\partial}\theta_X$. Defining the auxiliary function

$$F(g, X) = (n+1)2^{n+1} \int_M h_g \Delta_g \theta_X \omega_g^n,$$

it remains to show that this is independent of g. Note that F does not change if we replace θ_X by $\theta_X + c$ or h_g by $h_g + c$ for any constant c. In the following lemma, we will prove that F can be written without any terms involving h_g.

Lemma 3.2

$$\begin{aligned} F(g, X) = &\sum_{j=0}^{n} (-1)^j \frac{1}{j!(n-j)!} \int_M \Big((-\Delta_g\theta_X + \mathrm{Ric}(g) + (n-2j)(\theta_X + \omega_g))^{n+1} \\ &- (\Delta_g\theta_X - \mathrm{Ric}(g) + (n-2j)(\theta_X + \omega_g))^{n+1} \Big) \\ &- \mu 2^{n+1} \int_M (\theta_X + \omega_g)^{n+1} \end{aligned}$$

Proof. In order to prove this, we need the following binomial identities

$$\sum_{j=0}^{l} (-1)^j \binom{l}{j} (l-2j)^k = \begin{cases} 0 & \text{if } k < l \text{ or } k = l+1 \\ 2^l l! & \text{if } k = l \end{cases}$$

and the fact that $\int_M \Delta_g \theta_X \omega_g^n = 0$. Realizing that only degree $2n$ terms will contribute to the integral, by a straightforward computation, we can show that the right side of the above is equal to

$$(n+1)2^{n+1}\int_M (n\theta_X \operatorname{Ric}(g)\wedge\omega_g^{n-1} - \theta_X \mu\omega_g^n).$$

On the other hand, we can write

$$\begin{aligned}
F(g,X) &= (n+1)2^{n+1}\int_M \theta_X \Delta_g h_g \omega_g^n \\
&= (n+1)2^{n+1}\int_M \theta_X (s(g)-\mu)\omega_g^n \\
&= (n+1)2^{n+1}\int_M (n\theta_X \operatorname{Ric}(g)\wedge\omega_g^{n-1} - \theta_X \mu\omega_g^n).
\end{aligned}$$

This proves the lemma. □

We know that $\bar\partial\theta_X = -i_X\omega_g$ and we claim that

$$\bar\partial\Delta_g\theta_X = i_X \operatorname{Ric}(g).$$

This follows from

$$\begin{aligned}
i_X \operatorname{Ric}(g) &= -X^i \frac{\partial^2}{\partial z_i \partial\bar z_j}\log\det(g_{k\bar l})d\bar z_j \\
&= -\bar\partial\left(X^i\frac{\partial}{\partial z_i}\log\det(g_{k\bar l})\right) \\
&= -\bar\partial\left(X^i g^{k\bar l}\frac{\partial g_{k\bar l}}{\partial z_i}\right) \\
&= -\bar\partial\left(X^i g^{k\bar l}\frac{\partial g_{i\bar l}}{\partial z_k}\right) \\
&= -\bar\partial\left(g^{k\bar l}\frac{\partial}{\partial z_k}(X^i g_{i\bar l}) - g^{k\bar l}g_{i\bar l}\frac{\partial X^i}{\partial z_k}\right) \\
&= -\bar\partial\left(g^{k\bar l}\frac{\partial}{\partial z_k}(X^i g_{i\bar l})\right) \\
&= \bar\partial\left(g^{k\bar l}\frac{\partial}{\partial z_k}\frac{\partial}{\partial\bar z_l}\theta_X\right) \\
&= \bar\partial\Delta_g\theta_X.
\end{aligned}$$

Here the second line follows from the first because X^i is holomorphic and the fourth line follows from the third because ω_g is closed.

Now we are ready to show that the Calabi-Futaki invariant f_M is independent of the particular representative ω_g chosen in $[\omega]$.

Since the space of Kähler metrics is simply connected, it suffices to show that $\frac{\partial F}{\partial t}(g_t, X)|_{t=0} = 0$ for any family of metrics $\{g_t\}$ in a particular Kähler class.

Now we have that

$$\begin{aligned}\bar{\partial}\theta_{X,t} &= -i_X\omega_{g_t} \\ &= -i_X(\omega_g + \partial\bar{\partial}\phi_t) \\ &= \bar{\partial}\theta_{X,0} - \bar{\partial}(X(\phi_t))\end{aligned}$$

and therefore we can write

$$\theta_{X,t} = \theta_{X,0} - X(\phi_t),$$

because of the remark we made earlier about adding constants to θ. To simplify notations, we introduce

$$\begin{aligned}\psi_{X,t} &= -\Delta_{g_t}\theta_{X,t} + (n-2j)\theta_{X,t}, \\ \tilde{\psi}_{x,t} &= \Delta_{g_t}\theta_{X,t} + (n-2j)\theta_{X,t} \\ R(g_t) &= \mathrm{Ric}(g_t) + (n-2j)\omega_{g_t}, \\ \phi(x) &= x^{n+1},\end{aligned}$$

which yields

$$\begin{aligned}F(g_t, X) &= (n+1)2^{n+1}\int_M \theta_{X,t}\Delta_{g_t}h_{g_t}\omega_{g_t}^n \\ &= \sum_{j=0}^{n}(-1)^j\frac{1}{j!(n-j)!}\int_M \phi(\psi_{X,t} + R(g_t)) \\ &- \sum_{j=0}^{n}(-1)^j\frac{1}{j!(n-j)!}\int_M \phi(\tilde{\psi}_{X,t} - Ric(g_t) + (n-2j)\omega_{g_t}) \\ &- \mu\, 2^{n+1}\int_M \phi(\theta_{X,t} + \omega_{g_t}).\end{aligned}$$

We will only do the calculations for the first term because the calculations for other terms are the same. We want to show that F is independent of t and in order to deduce this, we must differentiate with respect to t, so

$$\frac{\partial F}{\partial t}(g_t, X) = \sum_{j=0}^{n}(-1)^j\frac{n+1}{j!(n-j)!}\int_M \phi(\dot{\psi}_{X,t} + \dot{R}(g_t), \psi_{X,t} + R(g_t), \cdots) + \cdots.$$

Here we have used that ϕ is a symmetric function. From the previous calculations, we know that

$$\bar{\partial}\psi_{X,t} = -i_X R(g_t)$$

and we can define α_t by

$$\dot{R}(g_t) = \bar{\partial}\alpha_t.$$

Then we have

$$\bar{\partial}\dot{\psi}_{X,t} = -i_X \dot{R}(g_t) = -i_X \bar{\partial}\alpha_t = -\bar{\partial}(i_X \alpha_t),$$

because X is holomorphic. So we can deduce that $\dot{\psi}_{X,t} = -i_X\alpha_t$, again because of the remark made earlier that adding constants has no effect on F. Therefore,

$$\begin{aligned}
&\int_M \phi(\dot{\psi}_{X,t} + \dot{R}(g_t), \psi_{X,t} + R(g_t), \ldots, \psi_{X,t} + R(g_t)) \\
&= \int_M \phi(-i_X\alpha_t + \bar{\partial}\alpha_t, \psi_{X,t} + R(g_t), \ldots, \psi_{X,t} + R(g_t)) \\
&= \int_M \phi(-i_X\alpha_t, \psi_{X,t} + R(g_t), \ldots, \psi_{X,t} + R(g_t)) \\
&\quad - \left(\cdots + \phi(\alpha_t, \psi_{X,t} + R(g_t), \ldots, \bar{\partial}(\psi_{X,t} + R(g_t)), \ldots, \psi_{X,t} + R(g_t)) + \cdots.\right) \\
&= -\int_M i_X \phi(\alpha_t, \psi_{X,t} + R(g_t), \ldots, \psi_{X,t} + R(g_t)).
\end{aligned}$$

To go from the one but last line to the last one we have used that $\bar{\partial}R(g_t) = 0$ and that

$$\bar{\partial}\psi_{X,t} = -i_X R(g_t) = -i_X(\psi_{X,t} + R(g_t)).$$

Now ϕ is a polynomial of degree $n+1$, so

$$\eta = \phi(\alpha_t, \psi_{X,t} + R(g_t), \ldots, \psi_{X,t} + R(g_t))$$

can be written as $\gamma_0 + \cdots + \gamma_{2n}$ and $i_X\eta = \beta_0 + \beta_1 + \cdots$. The only term in the above integral that will not give zero is the β_{2n}, but $\beta_{2n} = i_X\gamma_{2n+1}$ and $\gamma_{2n+1} = 0$. Therefore, this integral is zero. So are the integrals of the other terms.

Thus we have proved the following theorem, which is due to Calabi and Futaki

Theorem 3.3 *The integral*

$$\int_M X(h_g)\omega_g^n$$

is independent of g and defines a holomorphic invariant

$$f_M : Ka(M) \times \eta(M) \to \mathbb{C}.$$

In particular, there is a Kähler metric in $[\omega]$ *with constant scalar curvature only if* $f_M([\omega], -) = 0$.

Proof. We only need to prove the last statement, but this is clear because constant scalar curvature implies that $h_g =$ constant and hence $f_M = 0$ □

Remark 3.4 The proof of Theorem 3.3 here is different from the original ones by either Calabi or Futaki. This proof here is not simpler, however, it is more general, for example, let E be a hermitian bundle with a metric h over M and let ϕ be a symmetric polynomial of degree $n+1$. Then

$$\int \phi(\theta_X + R(h), \ldots, \theta_X + R(h))$$

is independent of h if $\bar{\partial}\theta_X = -i_X R(h)$, where $R(h)$ denotes the curvature form of h. To see this, one needs the Bott-Chern classes and a good reference is [2].

Corollary 3.5 *If* $f_M([\omega], -) = 0$, *then any extremal Kähler metric* g *in* $[\omega]$ *has constant scalar curvature.*

Proof. If g is extremal, then $X = g^{i\bar{j}} s(g)_{\bar{j}} \frac{\partial}{\partial z_i}$ is a holomorphic vector field. So we get

$$\begin{aligned} f_M([\omega], X) &= \int_M g^{i\bar{j}} s(g)_{\bar{j}} \frac{\partial h_g}{\partial z_i} \omega_g^n \\ &= \int_M g^{i\bar{j}} (s(g) - \mu)_{\bar{j}} \frac{\partial h_g}{\partial z_i} \omega_g^n \\ &= -\int_M (s(g) - \mu) \Delta h_g \omega_g^n \\ &= -\int_M (s(g) - \mu)^2 \omega_g^n \end{aligned}$$

and therefore $s(g) = \mu$. □

Corollary 3.6 *If* $X = [Y, Z] \in [\eta(M), \eta(M)]$, *then* $f_M([\omega], X) = 0$ *and therefore*

$$f_M : Ka(M) \times \eta(M)/[\eta(M), \eta(M)] \to \mathbb{C}.$$

Proof. $\mathrm{Re}(Z)$ generates a one parameter family of holomorphic transformations ϕ_t so that $\phi_t^* g$ is still a Kähler metric that lies in the same class as g. Hence we have that $F(\phi_t^* g, Y) = F(g, Y)$ and differentiating this with respect to t, we see that $F(g, [Y, \mathrm{Re}(Z)]) = 0$ and similarly $F(g, [Y, \mathrm{Im}(Z)]) = 0$. This proves the claim. □

3.2 Localization formula for Calabi-Futaki invariants

In this section, we state without proof a localization formula proved in [11]. The proof of this formula is similar to that of Bott's residue formula for Chern numbers. We will also compute the Calabi-Futaki invariants for three explicit Kähler manifolds.

Definition 3.7 *A holomorphic vector field X is non-degenerate if the zero set of X is the disjoint union of smooth connected complex submanifolds $\{Z_\lambda\}_{\lambda\in\Lambda}$ and if at each $z \in Z_\lambda$, the linear map*

$$DX : T_zM/T_zZ_\lambda \to T_zM/T_zZ_\lambda$$

is non-degenerate, that is,

$$\det(DX|_{T_zM/T_zZ_\lambda}) \neq 0.$$

In the case that $Z_\lambda = \{z\}$ and $X = X^i \frac{\partial}{\partial z_i}$, we have

$$DX(z) = \left(\frac{\partial X^i}{\partial z_j}\right)_{1\leq i,j\leq n} (z)$$

and so the non-degeneracy here means that

$$\det\left(\frac{\partial X^i}{\partial z_j}\right)_{1\leq i,j\leq n} (z) \neq 0.$$

Furthermore, if we are given a Kähler metric ω_g, we can make the following identification

$$T_zM/T_zZ_\lambda \simeq N_{M|Z_\lambda},$$

where $N_{M|Z_\lambda}$ is the normal bundle to Z_λ with respect to the metric ω_g, and hence

$$DX|_{T_zM/T_zZ_\lambda} = (\nabla X)^\perp|_{N_{M|Z_\lambda}}.$$

Given any Kähler class $\Omega \in Ka(M)$, we will define a "trace"

$$\mathrm{tr}_\Omega(X) : \{Z_\lambda\} \to \mathbb{C}$$

which is only well defined up to addition of constants. Fixing ω_g with $[\omega_g] = \Omega$, we may, as above, assume that $i_X\omega_g = -\bar\partial\theta_X$ and we set

$$\mathrm{tr}_\Omega(X)(Z_\lambda) = \theta_X(Z_\lambda).$$

Because $X|_{Z_\lambda} = 0$, we see $\bar\partial\theta_X = 0$ and hence $\theta_X|_{Z_\lambda}$ is constant. For a different metric $\omega_{g'}$ in the same Kähler class Ω, we have $\omega_{g'} = \omega_g + \partial\bar\partial\phi$, so

$$\theta'_X = \theta_X - X(\phi) + c,$$

where c is a constant independent of λ. It follows that

$$\theta'_X(Z_\lambda) = \theta_X(Z_\lambda) + c,$$

since X vanishes on Z_λ. Therefore, the trace is well defined up to addition of constants.

In the special case that $\Omega = \pi c_1(M)$, the form ω_g is the curvature of a hermitian metric on the line bundle $L = \Lambda^n T^{1,0}M$ and we have the induced vector field X^* on $\Lambda^n T^{1,0}M$, which is the canonical lifting of X on M.

At $z \in Z_\lambda$, $L_z = z \times \mathbb{C}$. If we denote by ξ the coordinate on $\mathbb{C}$, then we can write the vector field X^* as

$$X^* = a\frac{\partial}{\partial \xi},$$

where $a = \operatorname{tr}(DX)(z)$. So in this case, we have

$$\operatorname{tr}_{c_1(M)}(X)(z) = \operatorname{tr}(DX)(z) = \sum \frac{\partial X^i}{\partial z_i}(z).$$

Theorem 3.8 *For any $(\Omega, X) \in Ka(M) \times \eta(M)$ with X being non-degenerate, we have*

$$f_M(\Omega, X) = \pi^n$$

$$\sum_{\lambda\in\Lambda} \int_{Z_\lambda} \frac{(\operatorname{tr}(L_\lambda(X)) + c_1(M))(\operatorname{tr}_\Omega(X)(Z_\lambda) + \frac{1}{\pi}\Omega)^n - \frac{n\mu}{n+1}(\operatorname{tr}_\Omega(X)(Z_\lambda) + \frac{1}{\pi}\Omega)^{n+1}}{\det(L_\lambda(X) + \frac{\sqrt{-1}}{2\pi}K_\lambda)}.$$

Here $L_\lambda(X) = (\nabla X)^\perp|_{M|Z_\lambda}$ and K_λ is the curvature form of the induced metric on $N|_{M|Z_\lambda}$ by ω_g.

If $\Omega = \pi c_1(M)$, then this theorem is due to Futaki [10] and reads

$$f_M(c_1(M), X) = \frac{\pi^n}{n+1} \sum_{\lambda\in\Lambda} \int_{Z_\lambda} \frac{\operatorname{tr}(L_\lambda(X) + c_1(M))^{n+1}}{\det(L_\lambda(X) + \frac{\sqrt{-1}}{2\pi}K_\lambda)}.$$

We first of all need to explain why this expression is well defined. For simplicity, set for the moment, $L_\lambda(X) = x$, $c_1(M) = y$ and $\frac{\sqrt{-1}}{2\pi}K_\lambda = u$, then the above yields

$$f_M = \frac{\pi^n}{n+1} \sum_{\lambda\in\Lambda} \int_{Z_\lambda} F(x, y, u),$$

where F is of the form $\frac{f(x,y)}{g(x,u)}$. Here both f and g are analytic functions, so we can write down a power series for F,

$$F(x, y, u) = \sum a_{ijk} x^i y^j u^k.$$

This is an infinite series, but because only terms of a certain degree contribute to the integral, this becomes a finite expression.

We will not prove the theorem (see [22] for the proof), but we will consider two special cases and then compute the Calabi-Futaki invariant for certain examples.

Case 1 Assume that X only has isolated zeroes, so dim $(Z_\lambda) = 0$ and hence the only terms that will contribute to the integral are the degree 0 terms and therefore

$$f_M(\Omega, X) = \pi^n \sum_\lambda \frac{\operatorname{tr}(L_\lambda(X)) \operatorname{tr}_\Omega(X)^n - \frac{n\mu}{n+1} \operatorname{tr}_\Omega(X)^{n+1}}{\det(L_\lambda(X))},$$

and if $\Omega = \pi c_1(M)$, then this reduces to

$$f_M(c_1(M), X) = \frac{\pi^n}{n+1} \sum_\lambda \frac{\operatorname{tr}(L_\lambda(X))^{n+1}}{\det(L_\lambda(X))}.$$

In particular, if M has a Kähler-Einstein metric, then we know that $f_M = 0$ and therefore this puts a constraint on the zero set of any holomorphic vector fields.

Case 2 Assume that M is a complex surface (dim $M = 2$), then $\Lambda = \Lambda_0 \cup \Lambda_1$, where $\Lambda_i = \{\lambda \in \Lambda; \dim Z_\lambda = i\}$. Assume that $\Omega = \pi c_1(M)$, then

$$f_M(c_1(M), X) = \frac{\pi^2}{3} \sum_{\lambda \in \Lambda_0} \frac{\operatorname{tr}(L_\lambda(X))^3}{\det(L_\lambda(X))} + \frac{\pi^2}{3} \sum_{\lambda \in \Lambda_1} L_\lambda(X)(2c_1(M)(Z_\lambda) + 2 - 2g(Z_\lambda)).$$

To see this, we will consider the two terms separately. The first term follows from the case above. To understand the second term, we must do a little bit more work. Because Λ_1 consists of 1-dimensional submanifolds, we can omit the trace and determinant, so we have

$$\begin{aligned}
&\int_{Z_\lambda} \frac{\operatorname{tr}(L_\lambda(X) + c_1(M))^3}{\det(L_\lambda(X) + \frac{\sqrt{-1}}{2\pi} K_\lambda)} \\
&= \int_{Z_\lambda} \frac{(L_\lambda(X) + c_1(M))^3}{(L_\lambda(X) + \frac{\sqrt{-1}}{2\pi} K_\lambda)} \\
&= \int_{Z_\lambda} \frac{L_\lambda(X)^3 + 3c_1(M) L_\lambda(X)^2}{L_\lambda(X)(1 + \frac{\sqrt{-1}}{2\pi} K_\lambda L_\lambda(X)^{-1})} \\
&= \int_{Z_\lambda} \Big(L_\lambda(X)^2 + 3c_1(M) L_\lambda(X)\Big)\Big(1 - \frac{\sqrt{-1}}{2\pi} K_\lambda L_\lambda(X)^{-1}\Big) \\
&= \int_{Z_\lambda} \Big(3c_1(M) L_\lambda(X) - \frac{\sqrt{-1}}{2\pi} K_\lambda L_\lambda(X)\Big)
\end{aligned}$$

$$= L_\lambda(X)\Big(3c_1(M)(Z_\lambda) - \chi(N|_{M|Z_\lambda})\Big)$$
$$= L_\lambda(X)\Big(2c_1(M)(Z_\lambda) + 2 - 2g(Z_\lambda)\Big),$$

where we have omitted all terms that are not of a degree so that they can contribute to the integral. Furthermore, we used the fact that

$$\int_{Z_\lambda} \frac{\sqrt{-1}}{2\pi} K_\lambda = \chi(N|_{M|Z_\lambda}) = \chi(TM|_{Z_\lambda}) - \chi(Z_\lambda) = c_1(M)(Z_\lambda) + 2 - 2g(Z_\lambda).$$

Example 3.9 Let $M = \mathbb{C}P^n$, then $\mathrm{Aut}(M) = SL(n+1,\mathbb{C})/\sim$, where $A \sim \lambda A$ for some non-zero complex number λ, then

$$\eta(M) = \mathrm{Lie}(\mathrm{Aut}(M)) = sl(n+1,\mathbb{C}) = \{A \in \mathbb{C}^{n+1\times n+1}; \mathrm{tr}\, A = 0\}.$$

Because any $A \in sl(n+1,\mathbb{C})$ can be written as $A = BC - CB$, we have that $\eta(M) = [\eta(M), \eta(M)]$ and so from Corollary 3.5, we know that $f_M = 0$.

Example 3.10 Let M be the blow up of $\mathbb{C}P^2$ in a point. Without loss of generality, we may assume that we blow up in $[1:0:0]$ because we can always move points around by the automorphism group of $\mathbb{C}P^2$. Topologically, we see that $M \sim \mathbb{C}P^2 \# \overline{\mathbb{C}P^2}$, where the bar denotes the reversed orientation. Now $\mathbb{C}P^2 \setminus \{[1:0:0]\} = M \setminus E$ and in local coordinates, we get

$$M = \mathbb{C}P^2 \setminus \{[1:0:0]\} \cup \{[\xi:\eta] \times [1:x:y]; \xi y = \eta x\},$$

where $E = \{[\xi:\eta] \times [1:0:0]\} \simeq \mathbb{C}P^1$ is the exceptional divisor. Note that E has self intersection $E \cap E = -1$.

Claim $\eta(M) = \{X \in \eta(\mathbb{C}P^2); X([1:0:0]) = 0\}$.

Proof. Let X be a vector field on $\mathbb{C}P^2$ which vanishes at $[1:0:0]$. Then X induces a one-parameter subgroup ϕ_t on $\mathbb{C}P^2$ which fixes $[1:0:0]$. Clearly ϕ_t lifts to a one-parameter subgroup on M so that $\frac{d}{dt}\phi_t$ defines a holomorphic vector field Y on M. Conversely, let Y be a holomorphic vector field on M and ϕ_t be its integral flow of holomorphic automorphisms of M. Since E has self intersection -1 and other holomorphic curves always have positive self intersection numbers, the flow ϕ_t must fix E and hence can be descended to a flow of automorphisms of $\mathbb{C}P^2$, these automorphisms must fix $[1:0:0]$. Then the derivative of ϕ_t along t gives rise to the required vector field on $\mathbb{C}P^2$. This vector field vanishes at $[1:0:0]$. □

We will now consider the flow $\phi_t : [1:x:y] \to [1:e^t x:e^t y]$ which is defined on U_0. We can define it on all of $\mathbb{C}P^2$ by taking the limit of $\phi_t([1:\lambda x:\lambda y])$ as $\lambda \to \infty$. Since ϕ_t fix $[1:0:0]$, they lift to a one-parameter subgroup of automorphisms of M, still denoted by ϕ_t. One can then see that

$$\mathrm{Fix}(\phi_t) = \mathbb{C}P^1_\infty \cup E,$$

where $\mathbb{C}P^1_\infty = \{[0 : x : y]\}$. Let X be the vector field associated to ϕ_t on M. Then the zero set of X is $\mathbb{C}P^1_\infty \cup E$, which is the union of two one-dimensional submanifolds. It is not hard to see that $L_\lambda(X) = -1$ on $\mathbb{C}P^1_\infty$ and $L_\lambda(X) = 1$ on E (ϕ_t flows out of E and towards infinity). Therefore, we can now read off that

$$\begin{aligned}\frac{1}{\pi^2} f_M(c_1(M), X) &= \frac{1}{3}(2c_1(M)(E) + 2 - 2c_1(M)(\mathbb{C}P^1_\infty) - 2)\\ &= \frac{2}{3}(c_1(M)(E) - c_1(M)(\mathbb{C}P^1_\infty))\\ &= \frac{2}{3}(\chi(E) + E \cap E - \chi(\mathbb{C}P^1_\infty) - \mathbb{C}P^1_\infty \cap \mathbb{C}P^1_\infty)\\ &= \frac{2}{3}(2 - 1 - 2 - 1) = -\frac{4}{3},\end{aligned}$$

so $f_M \neq 0$ and therefore M does not admit a Kähler-Einstein metric.

Example 3.11 Let $M = \mathbb{C}P^2 \# 2\overline{\mathbb{C}P^2}$ be the blow up of $\mathbb{C}P^2$ in two points. Again without loss of generality, we may assume that we have blown up in $[1 : 0 : 0]$ and $[0 : 1 : 0]$ with exceptional divisors E_1 and E_2, respectively. Analogously to the above example, we have that

$$\eta(M) = \{X \in \eta(\mathbb{C}P^2); X([1 : 0 : 0]) = X([0 : 1 : 0]) = 0\}.$$

We will consider the same flow on $\mathbb{C}P^2$ as that in the last example. Let X be its associated vector field on M.

Let l be the line through $[1 : 0 : 0]$ and $[0 : 1 : 0]$, then l is preserved by the flow. The picture on the following page shows that

$$\mathrm{Fix}(\phi_t) = E_1 \cup F \cup \{l \cap E_2\},$$

where F is the image of $\mathbb{C}P^1_\infty$ after blowing up in $[0 : 1 : 0]$.

Now at $\{l \cap E_2\}$, we have

$$L_\lambda(X) = \mathrm{tr}\begin{pmatrix} 1 & 0 \\ 0 & -1 \end{pmatrix} = 0.$$

So again we can read off that

$$\begin{aligned}\frac{1}{\pi^2} f_M(c_1(M), X) &= \frac{1}{3}(2c_1(M)(E_1) - 2c_1(M)(F))\\ &= \frac{2}{3}(2 - 1 - (2 - 0)) = -\frac{2}{3},\end{aligned}$$

because $F \cap F = 0$ as blowing up $\mathbb{C}P^1$ in a point reduces the intersection number by one. And once more we can deduce that $f_M \neq 0$, so M does not admit a Kähler-Einstein metric.

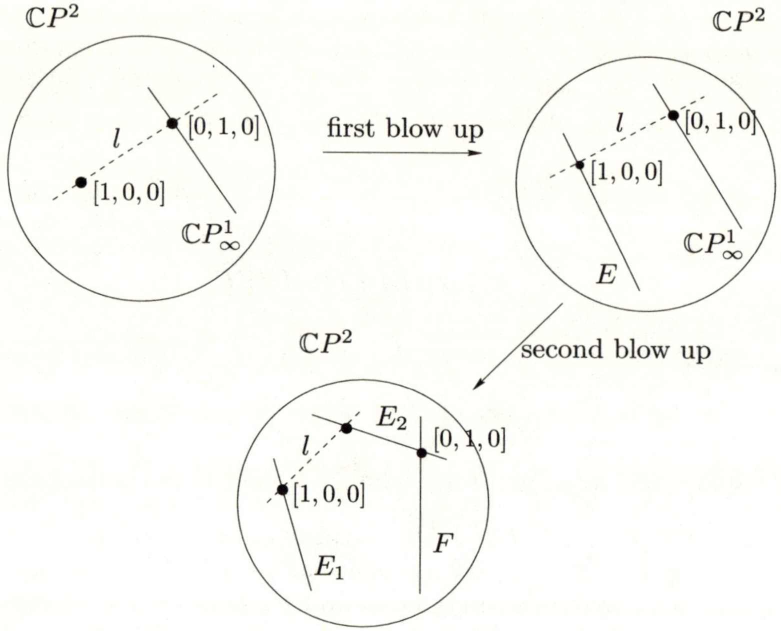

Remark 3.12 Note that $c_1(M) > 0$ if M is the blow up of $\mathbb{C}P^2$ in m points in general position for $m \leq 8$. If $m \geq 4$, we have that $\eta(M) = \{0\}$ and if $m = 3$, $\eta(M) \simeq \mathbb{C}^3$ and one can check that $f_M = 0$. In fact, these M admit Kähler-Einstein metrics (see [23]).

Chapter 4

Scalar curvature as a moment map

Let (V, ω) be a simply connected symplectic manifold and let G be a group acting on V preserving the symplectic form. Let $\mathfrak{g}$ be the Lie algebra of G which consists of all left-invariant vector fields on G. Then any $v \in \mathfrak{g}$ induces a one-parameter subgroup $\{\phi_t\}$ of G. Since G acts on V, ϕ_t induces a vector field X_v on V. It is well known that there exists a map m, called moment map, $m : V \to \mathfrak{g}^*$, satisfying

- m is G-equivariant with respect to the co-adjoint action on $\mathfrak{g}^*$,
- for all $v \in \mathfrak{g}$ and all $u \in TV$, $\omega(u, X_v) = dm(u)(v)$.

Our goal in this section is to prove that the scalar curvature of Kähler metrics is a moment map with respect to $\mathrm{Symp}(M)$, the group of symplectomorphisms of a compact Kähler manifold M. I learned this from [6]. In [8], Fijiki and Schumacher also studied moduli spaces of Kähler manifolds as quotients by symplectic diffeomorphism groups. Our presentation here follows closely the discussion by S.K. Donaldson in [6]. We should point out that though we assume for simplicity that M is simply-connected and Kähler in the following discussions, all calculations are still valid with small modifications in the general case of symplectic manifolds. Also it is not necessary for the readers to go through this section in order to understand other sections.

Let M be a simply-connected, compact Kähler manifold and fix a Kähler form ω. Set

$$\mathcal{J} = \{J : TM \to TM; \omega(Ju, Jv) = \omega(u, v),\ \omega(u, Ju) > 0 \text{ for } u \neq 0\}$$

and let $\mathcal{J}_{\text{int}} \subset \mathcal{J}$ be defined as

$$\mathcal{J}_{\text{int}} = \{J \in \mathcal{J}; J \,\text{is integrable}\},$$

which are both infinite-dimensional manifolds once appropriate norms are given. However, we will skip these technical difficulties. Let G be the group of exact symplectomorphisms of ω, G acts on $\mathcal{J}$ by

$$(\phi, J) \mapsto \phi_* J \phi_*^{-1}$$

which still lies in $\mathcal{J}$. Note that for $\phi_t \in G$, we get that $\phi_t^* \omega = \omega$, so if X is the vector field induced by ϕ_t, then $\mathcal{L}_X \omega = 0$ and by the Cartan formula for the Lie derivative, this implies that $d(i_X \omega) = 0$. Therefore, we can find a function H_X such that $i_X \omega = dH_X$. By requiring that $\int_M H_X \omega^n = 0$, such an H_X is unique, so the Lie algebra of G is

$$\mathfrak{g} \simeq \{H : M \to \mathbb{R}; \int_M H\omega^n = 0\} = C_0^\infty(M).$$

Secondly, observe that

$$T_J\mathcal{J} = \{A : TM \to TM; AJ + JA = 0, \omega(JAu, v) + \omega(u, JAv) = 0\},$$

in other words, JA is g_J-symmetric, where g_J is the metric induced by J, that is

$$g_J(u, v) = \omega(u, Jv).$$

Given $A \in T_J\mathcal{J}$, we can define $\mu_A(u, v) = \omega(Au, v)$ and we see that μ_A is anti J-invariant, that is $\mu_A(J\mu, Jv) = -\mu(u, v)$, and symmetric. The tangent space $T_J\mathcal{J}$ can be characterized to be the space of all such μ_A's. Finally for the orbit $G(J) = \{\phi_* J \phi_*^{-1}; \phi \in G\}$, we have

$$T_J G(J) = \{\mathcal{L}_X J; X \in \mathfrak{g}\}$$

Let now $\mathcal{D}$ be the distribution given by

$$\mathcal{D}_J = \{\mathcal{L}_X J, \mathcal{L}_{JX} J; X \in \mathfrak{g}\}.$$

Remark 4.1 In what follows we will always assume that the complex structure is integrable. We can also prove everything for arbitrary almost complex structures (see [6]), but we then need to take the Nijenhuis tensor into account and hence for simplicity we assume integrability.

Claim $\mathcal{D}_J \subset T_J\mathcal{J}$.

Proof. It is clear that both $\mathcal{L}_X J$ and $\mathcal{L}_{JX} J$ satisfy $AJ + JA = 0$. Because $X \in \mathfrak{g}$ it is also clear that $\mathcal{L}_X J$ satisfies the second condition to lie in $T_J\mathcal{J}$, and hence it suffices to show that $\mathcal{L}_{JX} J$ does so, too. Now we have

$$\begin{aligned}
(\mathcal{L}_{JX} J)Y &= \mathcal{L}_{JX}(JY) - J\mathcal{L}_{JX} Y \\
&= [JX, JY] - J[JX, Y] \\
&= [X, Y] + J[X, JY] \\
&= \mathcal{L}_X Y + J\mathcal{L}_X(JY) \\
&= -\mathcal{L}_X(J^2 Y) + J\mathcal{L}_X(JY) \\
&= -(\mathcal{L}_X J)(JY).
\end{aligned}$$

It shows that $(\mathcal{L}_{JX}J)Y = -(\mathcal{L}_X J)(JY)$ and therefore, we have

$$\begin{aligned}\omega(J(\mathcal{L}_{JX}J)u, v) &+ \omega(u, J(\mathcal{L}_{JX}J)v)\\ &= -\omega(J(\mathcal{L}_X J)Ju, v) - \omega(u, J(\mathcal{L}_X J)Jv)\\ &= \omega((\mathcal{L}_X J)Ju, Jv) + \omega(Ju, (\mathcal{L}_X J)Jv)\\ &= 0,\end{aligned}$$

so $\mathcal{L}_{JX}J \in T_J\mathcal{J}$. □

Remark 4.2 This also shows that $\mathcal{D}_J$ is a holomorphic distribution, this means that it is invariant with respect to the almost complex structure $\tilde{J}$ on $T_J\mathcal{J}$ induced by J.

Proposition 4.3 *$\mathcal{D}$ is integrable.*

Proof. Fix $J \in \mathcal{J}_{\text{int}}$ and let $K = \{$Kähler metrics within a fixed Kähler class$\}$.

It suffices to construct an integral submanifold $\mathcal{J}$ of $\mathcal{D}$ through each Kähler metric ω in K. Any Kähler metric in K near ω is of the form

$$\omega_f = \omega - dJdf = \omega + 2\sqrt{-1}\partial\bar{\partial}f$$

for some small $f \in C_0^\infty(M)$.

We now define a bundle S over K as follows: The fiber S_f over ω_f consists of all exact symplectomorphisms of ω_f. Note that each ω_f is J-invariant and $S_0 = G$.

By a lemma of J. Moser, we can always find a diffeomorphism ψ_f such that $\psi_f^*\omega_f = \omega$, and so given $\phi \in S_f$ we have that $\psi_f^{-1}\phi\psi_f \in S_0$.

Define the function $I : C_0^\infty(M) \times S_0 \to \mathcal{J}$ which associates to the pair $(f, \phi) \in C_0^\infty(M) \times S_0$ the complex structure $\phi(\psi_f^{-1}(J))$. We must check that this is well defined

$$\begin{aligned}\omega(\psi_f^{-1}(J)u, &\psi_f^{-1}(J)v)\\ &= \omega((\psi_f^{-1})_*J(\psi_f)_*u, (\psi_f^{-1})_*J(\psi_f)_*v)\\ &= \omega_f(J(\psi_f)_*u, J(\psi_f)_*v)\\ &= \omega_f((\psi_f)_*u, (\psi_f)_*v)\\ &= \omega(u, v)\end{aligned}$$

and since ϕ is a symplectomorphism, this shows that I is well defined. At a given point $(f, \phi) \in C_0^\infty(M) \times S_0$, we will compute the derivative of I applied to (H_1, H_2). Here we use the exponential to identify the symplectomorphisms with the Hamiltonian functions

$$\begin{aligned}
DI|_{(f,\phi)}(H_1,H_2) &= \frac{d}{dt}\Big((\phi e^{tH_2})(\psi^{-1}_{f+tH_1}(J))\Big)_{t=0}\\
&= \frac{d}{dt}\Big((\phi e^{tH_2})_*(\psi^{-1}_{f+tH_1})_* J(\psi_{f+tH_1})_* e_*^{-tH_2}\phi_*^{-1})\Big)_{t=0}\\
&= \phi_*\Big(\frac{d}{dt}(\psi^{-1}_{f+tH_1})_* J(\psi_{f+tH_1})_*\Big)_{t=0}\phi_*^{-1}\\
&\quad + \phi_*\mathcal{L}_{X_{H_2}}(\psi^{-1}_{f*}J\psi_{f*})\phi_*^{-1}.
\end{aligned}$$

We will consider both terms separately. First note that

$$\phi_*\mathcal{L}_{X_{H_2}}(\psi^{-1}_{f*}J\psi_{f*})\phi_*^{-1} = \mathcal{L}_{\phi_*X_{H_2}}(\phi(\psi_f^{-1}(J))),$$

we now want to find a nice expression for $\phi_*X_{H_2}$, so

$$\begin{aligned}
i_{\phi_*X_{H_2}}\omega(Y) &= \omega(\phi_*X_{H_2},Y)\\
&= \phi^*\omega(X_{H_2},\phi_*^{-1}Y)\\
&= \omega(X_{H_2},\phi_*^{-1}Y)\\
&= dH_2(\phi_*^{-1}Y)\\
&= \phi_*^{-1}(Y)(H_2)\\
&= Y((\phi^*)^{-1}H_2)\\
&= d(H_2\circ\phi^{-1})(Y)
\end{aligned}$$

and therefore

$$\phi_*X_{H_2} = X_{H_2\circ\phi^{-1}}.$$

For the first term, we need to do a bit more work. Observe that

$$\begin{aligned}
\frac{d}{dt}(\psi^{-1}_{f+tH_1})_* J(\psi_{f+tH_1})_* &= \frac{d}{dt}(\psi^{-1}_{f+tH_1})_*\psi_{f*}(\psi^{-1}_{f*}J\psi_{f*})\psi^{-1}_{f*}(\psi_{f+tH_1})_*\\
&= \mathcal{L}_{X_1}(\psi^{-1}_{f*}J\psi_{f*}),
\end{aligned}$$

where X_1 is the vector field

$$\frac{d}{dt}\Big(\psi_f^{-1}\psi_{f+tH_1}\Big)_{t=0}.$$

Again we want to find a nice expression for X_1. This will show that the distribution is integrable. First of all

$$\begin{aligned}
(\psi_f^{-1}\psi_{f+tH_1})^*\omega &= \psi^*_{f+tH_1}\omega_f\\
&= \psi^*_{f+tH_1}(\omega_f - tdJdH_1 + tdJdH_1)\\
&= \omega + t\psi^*_{f+tH_1}dJdH_1.
\end{aligned}$$

Since we can ignore second order terms in t, we have

$$\mathcal{L}_{X_1}\omega = \psi_f^* dJdH_1 = d(\psi_f^{-1}(J))d\psi_f^* H_1.$$

As $\mathcal{L}_{X_1}\omega = d(i_{X_1}\omega)$ and since M is simply connected, we get that

$$i_{X_1}\omega = \psi_f^{-1}(J)d\psi_f^* H_1 + dh$$

for some h. So we see that on the one hand,

$$\omega(X_1 - X_h, Y) = g(X_1 - X_h, \psi_f^{-1}(J)Y),$$

where g is the metric corresponding to the complex structure $\psi_f^{-1}(J)$, on the other hand,

$$\begin{aligned}\omega(X_1 - X_h, Y) &= \psi_f^{-1}(J)(d\psi_f^* H_1)(Y)\\ &= d\psi_f^* H_1(\psi_f^{-1}(J)Y)\\ &= \psi_f^{-1}(J)Y(\psi_f^* H_1)\\ &= g(\nabla\psi_f^* H_1, \psi_f^{-1}(J)Y)\end{aligned}$$

and hence $X_1 = \nabla\psi_f^* H_1 + X_h$. It follows

$$DI|_{(f,\phi)}(H_1, H_2) = \phi_*\Big(\mathcal{L}_{\nabla\psi_f^* H_1 + X_h}(\psi_f^{-1}(J))\Big)\phi_*^{-1} + \mathcal{L}_{X_{H_2\circ\phi^{-1}}}(\phi(\psi_f^{-1}(J)))$$

which reads by the same argument as above

$$DI|_{(f,\phi)}(H_1, H_2) = \mathcal{L}_{\nabla(\psi_f^* H_1\circ\phi^{-1}) + X_{h\cdot\phi^{-1}}}\phi(\psi_f^{-1}(J)) + \mathcal{L}_{X_{H_2\circ\phi^{-1}}}(\phi(\psi_f^{-1}(J))).$$

This shows that

$$DI|_{(f,\phi)}(H_1, H_2) \in \mathcal{D}_{\phi(\psi_f^{-1}(J))}$$

and that DI is surjective, hence the distribution is integrable. □

Consider now the following two operators

$$P : C_0^\infty(M) \to T_J\mathcal{J},$$

$$Q : T_J\mathcal{J} \to C_0^\infty(M),$$

where P represents the infinitesimal action of $\mathfrak{g}$ on $\mathcal{J}_{\text{int}}$ and Q represents the derivative of the map that associates to $J \in \mathcal{J}_{\text{int}}$ the scalar curvature of the metric g_J induced by J. To achieve the goal that we stated at the beginning of this section, to relate the scalar curvature of a Kähler metric to the moment map, we will prove that the following two L^2-pairings are the same:

$$(P(H), J\mu) = (H, Q(\mu)).$$

Here $H \in C_0^\infty(M)$ and $\mu \in T_J\mathcal{J}$. The pairing on the left-hand side equals

$$\frac{1}{2}\int_M \langle P(H), \mu\rangle_{g_J}\omega^n,$$

where we have used the inner product defined by g_J, and the pairing on the right-hand side is the standard L^2 inner product on functions, that is

$$\int_M HQ(\mu)\omega^n.$$

In order to prove that the two pairings are the same, we will compute both.

Let us first of all consider P. Since P is linear in H, we can assume that H has small compact support and so we can do the calculation in a local coordinate chart. Choose local coordinates $x_1, \dots, x_{2n}$ compatible with J so that $J = (J_j^i)$, $g_J = (g_{ik}) = (\omega_{ij}J_k^j)$ and

$$P(H) = \left(s_{ij} = P(H)(\frac{\partial}{\partial x_i}, \frac{\partial}{\partial x_j})\right).$$

Then we have

$$\omega\left(X_H, \frac{\partial}{\partial x_j}\right) = \frac{\partial H}{\partial x_j},$$

on the other hand,

$$\omega\left(X_H, \frac{\partial}{\partial x_j}\right) = g_J\left(JX_H, \frac{\partial}{\partial x_j}\right) = g_{jk}(JX_H)^k,$$

so

$$X_H^i = J_k^i g^{jk}\frac{\partial H}{\partial x_j}.$$

Writing for the moment

$$\mathcal{L}_{X_H}J = B_j^i\frac{\partial}{\partial x_i}dx_j,$$

we see that

$$B_j^i\frac{\partial}{\partial x_i} = \mathcal{L}_{X_H}J\left(\frac{\partial}{\partial x_j}\right) = \mathcal{L}_{X_H}\left(J\frac{\partial}{\partial x_j}\right) - J\mathcal{L}_{X_H}\frac{\partial}{\partial x_j}$$

and so

$$B_j^i = -J_j^k J_l^i g^{lp}H_{pk} - g^{ip}H_{pj}.$$

Therefore,

$$P(H) = \left(-J_j^k H_{ik} - J_i^k H_{jk}\right)$$

so that the pairing on $T_J\mathcal{J}_{\text{int}}$ reads

$$(P(H), J\mu) = -\int H_{ij}\mu_{kl}g^{ik}g^{jl}\omega^n = -\int H_{ij}\mu_{kl}g^{ik}g^{jl}\omega^n.$$

Now we will compute Q. Since $Q(\mu)$ is a tensor, there are no problems with the choice of coordinates. Let J_t be a family of almost complex structures with $J_0 = J$ and $\frac{d}{dt}J_t|_{t=0} = \mu$. Let g_t be the metric induced by J_t. Then the Christoffel symbols of the metric g_t are as follows,

$$\Gamma^i_{t,jk} = \frac{1}{2}g_t^{il}\left(\frac{\partial g_{t,lj}}{\partial x_k} + \frac{\partial g_{t,lk}}{\partial x_j} - \frac{\partial g_{t,jk}}{\partial x_l}\right),$$

where (g_t^{ij}) is the inverse of $(g_{t,ij})$ and

$$g_{t,ij} = g_t\left(\frac{\partial}{\partial x_i}, \frac{\partial}{\partial x_i}\right).$$

At $p \in M$, we can assume that

$$g_{ij}(p) = \delta_{ij},\ dg_{ij}(p) = 0 \ \text{ and } \ J(p) = \begin{pmatrix} 0 & -I \\ I & 0 \end{pmatrix},$$

where $g = g_0$. Then $\Gamma^i_{t,jk}$ are of order t, and the curvature of g_t at p is given by

$$\begin{aligned} R^t_{ijkl}(p) &= g_t\left(\nabla_{\frac{\partial}{\partial x_i}}\nabla_{\frac{\partial}{\partial x_j}}\frac{\partial}{\partial x_k} - \nabla_{\frac{\partial}{\partial x_j}}\nabla_{\frac{\partial}{\partial x_i}}\frac{\partial}{\partial x_k}, \frac{\partial}{\partial x_l}\right) \\ &= g_{t,sl}\frac{\partial\Gamma^s_{t,jk}}{\partial x_i} - g_{t,sl}\frac{\partial\Gamma^s_{t,ik}}{\partial x_j} + \mathcal{O}(t^2) \\ &= \frac{\partial^2 g_{t,jk}}{\partial x_i\partial x_l} - \frac{\partial^2 g_{t,il}}{\partial x_j\partial x_k} + \mathcal{O}(t^2), \end{aligned}$$

so

$$s(g_t) = g_t^{ik}g_t^{jl}\left(\frac{\partial^2 g_{t,jk}}{\partial x_i\partial x_l} - \frac{\partial^2 g_{t,il}}{\partial x_j\partial x_k}\right) + \mathcal{O}(t^2).$$

Differentiating it on t, we get

$$\frac{d}{dt}s(g_t) = -2\mu_{ik}R_{ijkj} + \mu_{jk,kj} - \mu_{jj,kk}.$$

and $Q(\mu) = \frac{d}{dt}s(g_t)|_{t=0}$.

Claim $\mu_{ik}R_{ijkj} = 0$ and $\mu_{jj} = 0$.

Proof. The first expression reads $\mu_{ik}R_{ijkj} = \mu_{ik}\,\mathrm{Ric}_{ik}$ and since μ is anti J-invariant and Ric is J-invariant, we can deduce that $\langle\mu, \mathrm{Ric}\rangle = 0$. Furthermore, μ being anti J-invariant implies that the trace of μ vanishes, so $\mu_{jj} = 0$. □

The corollary tells us that $Q(\mu) = \mu_{jk,kj}$ and hence we have the following pairing

$$(H, Q(\mu)) = \int HQ(\mu)\omega^n = -\int H_{jk}\mu_{jk}\omega^n.$$

Comparing this to the above pairing for P, we have shown that the two pairings are equal and we have found a relation between scalar curvature and moment maps. To be more precise, we have shown that the Calabi functional equals the square of the norm of the moment map m for the symplectic group G defined by scalar curvature as above. So we can deduce the following

Corollary 4.4 *g_J is an extremal metric if and only if*

$$J \in m^{-1}\Big(\frac{c_1(M)\omega^{n-1}}{\omega^n}\Big)$$

and therefore

$$\{\textit{extremal metrics}\}/\textit{holomorphic isometries} = m^{-1}\Big(\frac{c_1(M)\omega^{n-1}}{\omega^n}\Big)/G.$$

Chapter 5

Kähler-Einstein metrics with non-positive scalar curvature

5.1 The Calabi-Yau Theorem

We have seen that the Ricci curvature represents the first Chern class. In this section, we will consider the converse problem, namely, given a Kähler class $[\omega] \in H^2(M, \mathbb{R}) \cap H^{1,1}(M, \mathbb{C})$ on a compact Kähler manifold M and any form Ω representing the first Chern class, can we find a metric $\omega \in [\omega]$ such that $\operatorname{Ric}(\omega) = \Omega$? This is known as the Calabi conjecture and it was solved by Yau in 1976. We will state it here as a theorem and refer to it as the Calabi-Yau Theorem.

Theorem 5.1 *(Calabi-Yau) Let M be a compact Kähler manifold and let $[\omega] \in H^2(M, \mathbb{R}) \cup H^{1,1}(M, \mathbb{C})$. Given any form Ω representing $\pi c_1(M)$, there exists a unique Kähler metric $\omega \in [\omega]$ such that* $\operatorname{Ric}(\omega) = \Omega$.

Before proving this theorem, we will first discuss some corollaries and an example.

Corollary 5.2 *Any compact Kähler manifold with $c_1(M)_{\mathbb{R}} = c_2(M)_{\mathbb{R}} = 0$ is flat (this means $M = \mathbb{C}^n/\Gamma$).*

Proof. The Calabi-Yau Theorem tells us that $c_1(M) = 0$ implies that there exists a Ricci-flat metric (that is $\operatorname{Ric}(\omega) = \Omega = 0$). By the Uniformization Theorem (2.13), we know that $\widetilde{M} = \mathbb{C}^n$ and therefore $M = \mathbb{C}^n/\Gamma$. □

This shows that the flatness is characterized by the first two Chern classes.

Corollary 5.3 *If $c_1(M) > 0$, then M has a Kähler metric with positive Ricci curvature (this implies that M is simply connected, see [13]).*

Proof. If $c_1(M) > 0$, then there is a form $\Omega > 0$ representing $\pi c_1(M)$ and by the Calabi-Yau Theorem, we can find ω such that $\Omega = \mathrm{Ric}(\omega) > 0$. □

Remark 5.4 It is unknown whether this result is also true for symplectic manifolds.

Corollary 5.5 *Assume that $c_1(M) = 0$. Given any Kähler class $[\omega]$, then*

$$\mathrm{Aut}(M, [\omega]) = \{\sigma : M \to M; \sigma \,\text{is biholomorphic and}\, \sigma^*[\omega] = [\omega]\}$$

is a finite-dimensional Lie group.

Proof. By the Calabi-Yau Theorem, there exists a unique Ricci-flat metric $\omega \in [\omega]$. Observe that $\sigma^*\omega$ is still Ricci-flat and is also Kähler (since σ is biholomorphic). By assumption

$$[\sigma^*\omega] = \sigma^*[\omega] = [\omega]$$

and therefore, by uniqueness,

$$\sigma^*\omega = \omega.$$

So σ is contained in the isometry group $\mathrm{Isom}(\omega)$ of ω and we have shown that $\mathrm{Aut}(M, [\omega]) = \mathrm{Isom}(\omega)$ and the latter one is a finite-dimensional Lie group (see [14]). □

Remark 5.6 There exists a compact $K3$ surface M (that is a complex surface with $\pi_1(M) = 0, c_1(M) = 0$) such that $\mathrm{Aut}(M)$ is an infinite discrete group. Hence, the extra assumption $\sigma^*[\omega] = [\omega]$ is necessary.

Example 5.7 Let $M \subset \mathbb{C}P^3$ be defined by a quartic (that is degree 4) homogeneous polynomial. Example 2.9 tells us that $c_1(M) = 0$ and it follows from the Lefschetz Hyperplane Theorem that $\pi_1(M) = 0$. The Calabi-Yau Theorem says that M has a Ricci-flat metric, but M is not flat (if M were flat then $M = \mathbb{C}^n/\Gamma$ and hence $\pi_1(M) = \Gamma$). A question that arises is whether it is possible to construct this metric in terms of classical functions.

We will now give the proof of the Calabi conjecture due to Yau (see [27]).

Proof. Choose any Kähler metric $\omega \in [\omega]$. For convenience, we will drop the normalization factor $\frac{\sqrt{-1}}{2}$ in ω. Then, in local complex coordinates $z_1, \ldots, z_n$,

$$\omega = g_{i\bar{j}} dz_i \wedge d\bar{z}_j$$

and

$$\mathrm{Ric}(\omega) = -\partial\bar{\partial} \log \det(g_{i\bar{j}}).$$

Both the Ricci curvature and Ω represent the first Chern class and therefore the $\partial\bar{\partial}$-lemma tells us that we can find f such that

$$\Omega - \mathrm{Ric}(\omega) = \partial\bar{\partial} f,$$

where f is unique after normalizing to

$$\int_M (e^f - 1)\omega^n = 0.$$

Note that f only depends on ω and Ω.

Again using the $\partial\bar{\partial}$-lemma, we know that any other metric in $[\omega]$ is of the form $\omega + \partial\bar{\partial}\phi$. Suppose that we have found the right function ϕ, so

$$\mathrm{Ric}(\omega + \partial\bar{\partial}\phi) = \Omega = \mathrm{Ric}(\omega) - \partial\bar{\partial} f.$$

This reads in local coordinates,

$$-\partial\bar{\partial} \log\det\left(g_{i\bar{j}} + \frac{\partial^2\phi}{\partial z_i \partial \bar{z}_j} \right) = -\partial\bar{\partial}\log\det(g_{i\bar{j}}) - \partial\bar{\partial} f.$$

Although this is only locally defined, the following is globally defined

$$\partial\bar{\partial}\log\left(\frac{\det(g_{i\bar{j}} + \frac{\partial^2\phi}{\partial z_i\partial\bar{z}_j})}{\det(g_{i\bar{j}})} \right) = \partial\bar{\partial} f.$$

Therefore

$$\frac{\det(g_{i\bar{j}} + \frac{\partial^2\phi}{\partial z_i\partial\bar{z}_j})}{\det(g_{i\bar{j}})} = e^{f+c},$$

which is equivalent to

$$(\omega + \partial\bar{\partial}\phi)^n = e^{f+c}\omega^n.$$

In order to determine c, observe that

$$\int_M e^{f+c}\omega^n = \int_M (\omega + \partial\bar{\partial}\phi)^n = \int_M \omega^n + \partial(\dots) = \int_M \omega^n,$$

which follows from the Stokes Theorem. This implies that $c = 0$.

So we have shown that $\mathrm{Ric}(\omega + \partial\bar{\partial}\phi) = \Omega$ implies

$$(\omega + \partial\bar{\partial}\phi)^n = e^f\omega^n,$$

which is known as the complex Monge-Ampère equation. In fact, it is true that

$$\mathrm{Ric}(\omega + \partial\bar{\partial}\phi) = \Omega \iff (\omega + \partial\bar{\partial}\phi)^n = e^f\omega^n.$$

Calabi proved in the 50's the uniqueness part of his conjecture using the Maximum Principle.

Assume that there are two metrics ω_1 and ω_2 with

$$\mathrm{Ric}(\omega_1) = \mathrm{Ric}(\omega_2) = \Omega,$$

then

$$\omega_1 = \omega + \partial\bar{\partial}\phi_1$$

and

$$\omega_2 = \omega + \partial\bar{\partial}\phi_2.$$

Without loss of generality we can assume that $\phi_2 = 0$, $\omega_2 = \omega$ and $\phi_1 = \phi$ so that we have

$$\begin{aligned} 0 &= \omega_2^n - \omega_1^n \\ &= \omega^n - (\omega + \partial\bar{\partial}\phi)^n \\ &= (\omega - (\omega + \partial\bar{\partial}\phi)) \wedge (\omega^{n-1} + \omega^{n-2} \wedge \omega_1 + \cdots + \omega \wedge \omega_1^{n-2} + \omega_1^{n-1}) \\ &= -\partial\bar{\partial}\phi \wedge (\omega^{n-1} + \omega^{n-2} \wedge \omega_1 + \cdots + \omega_1^{n-1}) \end{aligned}$$

and multiplying by ϕ and integrating, we see that

$$0 = -\int_M \phi\partial\bar{\partial}\phi \wedge (\omega^{n-1} + \cdots + \omega_1^{n-1}) = \int_M \partial\phi \wedge \bar{\partial}\phi \wedge (\omega^{n-1} + \cdots + \omega_1^{n-1}).$$

Now because ω_1 and ω_2 are Kähler metrics and $\omega_1, \omega_2 \geq 0$, we get that

$$\frac{1}{V}\,\partial\phi \wedge \bar{\partial}\phi \wedge (\omega^{n-1} + \cdots + \omega_1^{n-1}) \geq \frac{1}{V}\,\partial\phi \wedge \bar{\partial}\phi \wedge \omega^{n-1},$$

where $V = \int_M \omega^n$. Here we say that two top degree forms $u = f dz_1 \wedge \cdots \wedge d\bar{z}_n$ and $\tilde{u} = \tilde{f} dz_1 \wedge \cdots \wedge d\bar{z}_n$ satisfy $u \geq \tilde{u}$ if $f \geq \tilde{f}$. So we have

$$0 \geq \frac{1}{V}\int_M \partial\phi \wedge \bar{\partial}\phi \wedge \omega^{n-1} = \frac{1}{nV}\int |\partial\phi|^2\omega^n = \frac{1}{2nV}\int_M |\nabla\phi|^2\omega^n$$

and we can deduce that $\nabla\phi = 0$, so $\phi =$ constant and therefore $\omega_1 = \omega_2$.

Hence we have proved the uniqueness part of the theorem and we now consider the existence part, so we want to solve $(\omega + \partial\bar{\partial}\phi)^n = e^f\omega^n$ for ϕ. The proof will follow by the continuity method: Define $f_s = sf + c_s$ for c_s constants and $0 \leq s \leq 1$. Requiring that $\int_M (e^{f_s} - 1)\omega^n = 0$ determines the constants uniquely. Observe that $f_0 = 0$ and $f_1 = f$. Consider the following family of equations

$$(\omega + \partial\bar{\partial}\phi)^n = e^{f_s}\omega^n . \qquad (1_s)$$

The solution of (1_s) is unique up to constants. Define

$$S = \{s \in [0,1]; (1_t)\,\text{is solvable for}\, t \leq s\}.$$

To prove that (1_1) is solvable, it suffices to prove that S is non-empty, open and closed. Clearly, $0 \in S$ (set $\phi =$ constant).

We will first show that S is open. This means that assuming $s \in S$, we must solve (1_t) for t close to s. Let ϕ_s be the solution of (1_s), then

$$\omega_s = \omega + \partial\bar{\partial}\phi_s$$

and

$$\omega_s^n = e^{f_s}\omega^n.$$

Suppose that ϕ_t is what we look for, so

$$(\omega + \partial\bar{\partial}\phi_t)^n = e^{f_t}\omega^n,$$

then

$$(\omega_s + \partial\bar{\partial}(\phi_t - \phi_s))^n = e^{f_t - f_s}\omega_s^n.$$

Setting $\psi = \phi_t - \phi_s$, all that remains to do is to show that we can solve this for t close to s. Note that this is a perturbation problem and yields

$$\log \frac{(\omega_s + \partial\bar{\partial}\psi)^n}{\omega_s^n} = f_t - f_s.$$

Since ψ is small, we can expand the left-hand side and get

$$\begin{aligned} f_t - f_s &= \log(1 + \Delta_{\omega_s}\psi + Q(D^2\psi, D^2\psi) + \cdots) \\ &= \Delta_{\omega_s}\psi + Q(D^2\psi, D^2\psi) - \frac{1}{2}(\Delta_{\omega_s}\psi)^2 + \ldots \end{aligned}$$

where Q denotes the quadratic terms and the dots denote the higher order terms, $f_t - f_s$ is of order $t - s$.

Introduce

$$C^{k,\frac{1}{2}}(M,\mathbb{R}) = \{u : M \to \mathbb{R}; u\,\text{is}\,C^k\text{-smooth},\ \sup_{x \neq y} \frac{|D^k u(x) - D^k u(y)|}{d(x,y)^{\frac{1}{2}}} < \infty\},$$

$$\|v\|_{k,\frac{1}{2}} = \sum_{i=0}^{k} \sup_x |D^i v(x)| + \sup_{x\neq y} \frac{|D^k u(x) - D^k u(y)|}{d(x,y)^{\frac{1}{2}}}.$$

We denote by $C_0^{\cdot,\cdot}(M,\mathbb{R})$ the subspace consisting of all u in $C^{\cdot,\cdot}(M,\mathbb{R})$ with the extra linear condition

$$\int u\omega_s^n = 0,$$

whereas $C^{\cdot,\cdot}(M,\mathbb{R})_0$ denotes the subset of all u in $C^{\cdot,\cdot}(M,\mathbb{R})$ satisfying the extra non-linear condition

$$\int (e^u - 1)\omega_s^n = 0.$$

Note that

$$\int (e^{f_t - f_s} - 1)\omega_s^n = \int (e^{f_t - f_s} - 1)e^{f_s}\omega^n = \int ((e^{f_t} - 1) - (e^{f_s} - 1))\omega^n = 0$$

implies that

$$0 = \int \left(e^{\log \frac{(\omega_s + \partial\bar{\partial}\psi)^n}{\omega_s^n}} - 1 \right) \omega_s^n = \int (\omega_s + \partial\bar{\partial}\psi)^n - \omega_s^n.$$

By the above, it follows that we can define an operator

$$\Phi : C_0^{2,\frac{1}{2}}(M,\mathbb{R}) \to C^{0,\frac{1}{2}}(M,\mathbb{R})_0$$

by

$$\Phi(\psi) = \log \frac{(\omega_s + \partial\bar{\partial}\psi)^n}{\omega_s^n}.$$

This is well defined. We must prove that for $\|f_t - f_s\|_{0,\frac{1}{2}}$ sufficiently small, we can find a ψ with

$$\Phi(\psi) = f_t - f_s \quad \text{and} \quad \|\psi\|_{2,\frac{1}{2}} \le C\|f_t - f_s\|_{0,\frac{1}{2}}.$$

We will apply the Implicit Function Theorem and hence it suffices to check that

$$D\Phi|_{\psi=0} : C_0^{2,\frac{1}{2}}(M,\mathbb{R}) \to C_0^{0,\frac{1}{2}}(M,\mathbb{R})$$

is invertible (note that the tangent space to $C^{0,\frac{1}{2}}(M,\mathbb{R})_0$ at $u = 0$ is $C_0^{0,\frac{1}{2}}(M,\mathbb{R})$). Now

$$D\Phi|_{\psi=0}(u) = \Delta_{\omega_s} u$$

and it follows from [12] that

$$\Delta_{\omega_s} : C_0^{2,\frac{1}{2}}(M,\mathbb{R}) \to C_0^{0,\frac{1}{2}}(M,\mathbb{R})$$

is invertible. Therefore, $\Phi(\psi) = f_t - f_s$ is solvable for $|t - s|$ sufficiently small and so we have shown that S is open.

It remains to prove that S is closed, that is, suppose that we have a sequence $s_i \in S$ with $\lim_{i\to\infty} s_i = s_\infty$, then we must show that $s_\infty \in S$. Now the s_i correspond to solutions ϕ_i of (1_{s_i}) which are unique up to constants. We must prove that by taking a subsequence if necessary, there exist c_i such that $\phi_i - c_i$ converge to some ϕ_∞ in the $C^{2,\frac{1}{2}}$-topology. Since

$$(\omega + \partial\bar{\partial}(\phi_i - c_i))^n = e^{f_{s_i}}\omega^n,$$

we know that

$$(\omega + \partial\bar{\partial}\phi_\infty)^n = e^{f_{s_\infty}}\omega^n$$

and hence $s_\infty \in S$. By the Arzela-Ascoli lemma, it suffices to show the following a priori estimate $\|\phi_i - c_i\|_3 \leq C$. Let ϕ be any solution of $(*)_s$. We will establish an a priori estimate for $||\phi||_3$.

We will first state the following two facts that can be found in [1]

– Let g be the metric corresponding to $\omega = \omega_g$, then there exists a Green function $G(x,y)$ on $M \times M \setminus \Delta$ satisfying

$$0 \leq G(x,y) \leq \frac{c}{d(x,y)^{2n-2}}$$

and for all $\phi \in C^\infty(M)$,

$$\phi(x) = \frac{1}{V}\int_M \phi(y)\omega^n(y) - \frac{1}{V}\int_M \Delta_g\phi(y)G(x,y)\omega^n(y).$$

To see that we have positivity follows from the fact that we can always add a sufficiently large constant to G to ensure positivity.

– (Sobolev inequality) For a fixed metric, there exist constants $c_1, c_2 > 0$ depending on (M, ω_g) such that for all $f \in C^1(M, \mathbb{R})$

$$c_1\left(\frac{1}{V}\int_M |f|^{\frac{2n}{n-1}}\omega^n\right)^{\frac{n-1}{n}} - \frac{c_2}{V}\int_M |f|^2\omega^n \leq \frac{1}{V}\int_M |\nabla f|^2\omega^n.$$

This is true because on a compact manifold, we can use a partition of unity to cut the manifold in pieces which are diffeomorphic to a domain in Euclidean space and there we can apply the standard Sobolev inequality.

To show the desired estimate, we will proceed in three steps. In what follows we will always use C to denote a uniform constant, but this capital may mean many different constants and it should be clear from the context what is meant.

Step 1 (C^0-estimate) We know that ϕ is unique up to a constant, so we can choose the constant so that $\sup_M \phi = -1$ and because M is compact, we can assume that ϕ attains its supremum at x_0 so that we have

$$-1 = \phi(x_0) = -\frac{1}{V}\int_M |\phi(y)|\omega^n - \frac{1}{V}\int_M \Delta\phi(y)G(x_0,y)\omega^n,$$

where we have used that ϕ is always negative so $\phi = -|\phi|$. Since $\omega + \partial\bar{\partial}\phi$ determines a metric, by taking trace with respect to ω, we have

$$0 < n + \Delta\phi.$$

Substituting this in the above gives

$$-1 + \frac{1}{V}\int_M |\phi(y)|\omega^n = -\frac{1}{V}\int_M \Delta\phi(y)G(x_0,y)\omega^n \leq \frac{n}{V}\int_M G(x_0,y)\omega^n$$

and therefore we have proved

$$\frac{1}{V}\int_M |\phi(y)|\omega^n \le C.$$

Define $\phi_- = -\phi \ge 1$ and $\omega' = \omega - \partial\bar\partial\phi_-$, then we see as before

$$\begin{aligned}(e^{f_s}-1)\omega^n =&(\omega')^n - \omega^n\\ =&-\partial\bar\partial\phi_- \wedge ((\omega')^{n-1} + (\omega')^{n-2}\wedge\omega + \cdots + \omega'\wedge\omega^{n-2} + \omega^{n-1})\end{aligned}$$

where both ω and ω' are metrics so $\omega, \omega' > 0$. For any $p \ge 1$, we multiply it by ϕ_-^p, and integrate over M to get

$$\begin{aligned}&-\frac{1}{V}\int_M \phi_-^p \partial\bar\partial\phi_- \wedge ((\omega')^{n-1} + \cdots + \omega^{n-1})\\ &\qquad \ge \frac{1}{V}\int_M \partial\phi_-^p \wedge \bar\partial\phi_- \wedge \omega^n\\ &\qquad = \frac{p}{V}\int_M \phi_-^{p-1}\partial\phi_- \wedge \bar\partial\phi_- \wedge \omega^{n-1}\\ &\qquad = \frac{p}{V}\int_M \phi_-^{\frac{p-1}{2}}\partial\phi_- \wedge \phi_-^{\frac{p-1}{2}}\bar\partial\phi_- \wedge \omega^{n-1}\\ &\qquad = \frac{4p}{V(p+1)^2}\int_M \partial\phi_-^{\frac{p+1}{2}} \wedge \bar\partial\phi_-^{\frac{p+1}{2}}\omega^{n-1}\\ &\qquad = \frac{4p}{Vn(p+1)^2}\int_M |\nabla\phi_-^{\frac{p+1}{2}}|^2\omega^n\\ &\qquad \ge \frac{4p}{n(p+1)^2}\left(c_1\left(\frac{1}{V}\int_M |\phi_-^{\frac{p+1}{2}}|^{\frac{2n}{n-1}}\omega^n\right)^{\frac{n-1}{n}} - \frac{c_2}{V}\int_M |\phi_-^{\frac{p+1}{2}}|^2\omega^n\right).\end{aligned}$$

On the other hand, we have that

$$\begin{aligned}-\frac{1}{V}\int_M \phi_-^p \partial\bar\partial\phi_- \wedge ((\omega')^{n-1} + \cdots + \omega^{n-1}) &= \frac{1}{V}\int_M \phi_-^p(e^{f_s}-1)\omega^n\\ &\le \frac{c}{V}\int_M \phi_-^p\omega^n\\ &\le \frac{c}{V}\int_M \phi_-^{p+1}\omega^n,\end{aligned}$$

since the f_s are uniformly bounded (this bound is independent of s) and $\phi_- \ge 1$. Putting all these together, we get

$$\frac{4p}{n(p+1)^2}\left(c_1\left(\frac{1}{V}\int_M |\phi_-^{\frac{p+1}{2}}|^{\frac{2n}{n-1}}\omega^n\right)^{\frac{n-1}{n}} - \frac{c_2}{V}\int_M |\phi_-^{\frac{p+1}{2}}|^2\omega^n\right) \le \frac{c}{V}\int_M |\phi_-^{p+1}|\omega^n,$$

which after some rearranging reads

$$\left(\frac{1}{V}\int_M |\phi_-^{p+1}|^{\frac{n}{n-1}}\omega^n\right)^{\frac{n-1}{n}} \leq \frac{C(p+1)}{V}\int_M |\phi_-^{p+1}|\omega^n.$$

By raising both sides to the power $\frac{1}{p+1}$, we obtain

$$\|\phi_-\|_{L^{(p+1)\frac{n}{n-1}}} \leq (C(p+1))^{\frac{1}{p+1}}\|\phi_-\|_{L^{p+1}}.$$

We will apply Moser's iteration method. Choose $p_0 = 1$ and define p_i by

$$p_i + 1 = \frac{n}{n-1}(p_{i-1}+1),$$

then

$$\|\phi_-\|_{L^{p_i+1}} \leq \prod_{j=0}^{i-1}(C(p_j+1))^{\frac{1}{p_j+1}}\|\phi_-\|_{L^2}.$$

And if we now let $i \to \infty$, then $p_i + 1 \to \infty$, so

$$\sup_M |\phi_-| = \lim_{i\to\infty}\|\phi_-\|_{L^{p_i+1}} \leq \prod_{j=0}^{\infty}(C(p_j+1))^{\frac{1}{p_j+1}}\|\phi_-\|_{L^2} < \infty,$$

which we can see by considering

$$e^{\log\prod_{j=0}^{\infty}(C(p_j+1))^{\frac{1}{p_j+1}}} = e^{\sum_{j=0}^{\infty}\frac{1}{p_j+1}(\log C + \log(p_j+1))}.$$

So we have shown that

$$\sup_M |\phi| = \sup_M |\phi_-| \leq C\|\phi\|_{L^2}.$$

Furthermore, we have

$$\begin{aligned}\frac{c}{V}\int_M |\phi|\omega^n &\geq \frac{1}{V}\int_M \phi(1-e^{f_s})\omega^n\\ &= \frac{1}{V}\int_M \phi(\omega^n - (\omega+\partial\bar{\partial}\phi)^n)\\ &\geq \frac{1}{nV}\int_M |\nabla\phi|^2\omega^n\\ &\geq \frac{\lambda_1(\omega)}{nV}\left(\int_M |\phi|^2\omega^n - \left(\int_M \phi\omega^n\right)^2\right),\end{aligned}$$

where the last step follows from the Poincaré inequality, so we deduce from the above that

$$\|\phi\|_{L^2} \leq C(\|\phi\|_{L^1}+1)$$

and hence $\sup_M |\phi| \leq C$.

Step 2 (**C^2-estimate**) First observe that $\|\nabla^2\phi\|_{C^0} \le \max\{n+\Delta\phi, n\}$. To see this, note that the metric satisfies $g_{i\bar{j}} + \frac{\partial^2\phi}{\partial z_i \partial \bar{z}_j} > 0$ (positive definite, hence all the eigenvalues are positive) and therefore

$$\|g_{i\bar{j}} + \frac{\partial^2\phi}{\partial z_i\partial\bar{z}_j}\|_g \le \mathrm{tr}_g\left(g_{i\bar{j}} + \frac{\partial^2\phi}{\partial z_i\partial\bar{z}_j}\right) = n+\Delta\phi.$$

Our aim is now to find a bound for $\mathrm{tr}_\omega(\omega') = n + \Delta\phi$ where $\omega' = \omega + \partial\bar{\partial}\phi$. We will follow Yau's arguments [27] using the Maximum Principle.

Locally we write $\omega' = \{g'_{i\bar{j}}\}$ and $\omega = \{g_{i\bar{j}}\}$ and we set $f_s = F$ so that we have

$$(\omega + \partial\bar{\partial}\phi)^n = e^F\omega^n,$$

which reads

$$\log\det\left(g_{i\bar{j}} + \frac{\partial^2\phi}{\partial z_i\partial\bar{z}_j}\right) = F + \log\det(g_{i\bar{j}}).$$

Now differentiate both sides with respect to $\frac{\partial}{\partial z_k}$

$$(g')^{i\bar{j}}\left(\frac{\partial g_{i\bar{j}}}{\partial z_k} + \frac{\partial^3\phi}{\partial z_i\partial\bar{z}_j\partial z_k}\right) - g^{i\bar{j}}\frac{\partial g_{i\bar{j}}}{\partial z_k} = \frac{\partial F}{\partial z_k},$$

and differentiating again with respect to $\frac{\partial}{\partial \bar{z}_l}$ yields

$$\begin{aligned}&(g')^{i\bar{j}}\left(\frac{\partial^2 g_{i\bar{j}}}{\partial z_k\partial\bar{z}_l} + \frac{\partial^4\phi}{\partial z_i\partial\bar{z}_j\partial z_k\partial\bar{z}_l}\right) + g^{t\bar{j}}g^{i\bar{s}}\frac{\partial g_{t\bar{s}}}{\partial\bar{z}_l}\frac{\partial g_{i\bar{j}}}{\partial z_k} - g^{i\bar{j}}\frac{\partial^2 g_{i\bar{j}}}{\partial z_k\partial\bar{z}_l}\\ &-(g')^{t\bar{j}}(g')^{i\bar{s}}\left(\frac{\partial g_{t\bar{s}}}{\partial\bar{z}_l} + \frac{\partial^3\phi}{\partial z_t\partial\bar{z}_s\partial\bar{z}_l}\right)\left(\frac{\partial g_{i\bar{j}}}{\partial z_k} + \frac{\partial^3\phi}{\partial z_i\partial\bar{z}_j\partial z_k}\right) = \frac{\partial^2 F}{\partial z_k\partial\bar{z}_l}.\end{aligned}$$

Assume that we have normal coordinates at the given point, so $g_{i\bar{j}} = \delta_{ij}$ and that the first order derivatives of g vanish. Now taking the trace of both sides results in

$$\begin{aligned}\Delta F = {}& g^{k\bar{l}}(g')^{i\bar{j}}\left(\frac{\partial^2 g_{i\bar{j}}}{\partial z_k\partial\bar{z}_l} + \frac{\partial^4\phi}{\partial z_i\partial\bar{z}_j\partial z_k\partial\bar{z}_l}\right) - g^{k\bar{l}}(g')^{t\bar{j}}(g')^{i\bar{s}}\frac{\partial^3\phi}{\partial z_t\partial\bar{z}_s\partial\bar{z}_l}\frac{\partial^3\phi}{\partial z_i\partial\bar{z}_j\partial z_k}\\ &- g^{k\bar{l}}g^{i\bar{j}}\frac{\partial^2 g_{i\bar{j}}}{\partial z_k\partial\bar{z}_l}.\end{aligned}$$

On the other hand, we also have

$$\begin{aligned}\Delta'(\Delta\phi) &= (g')^{k\bar{l}}\frac{\partial^2}{\partial z_k\partial\bar{z}_l}\left(g^{i\bar{j}}\frac{\partial^2\phi}{\partial z_i\partial\bar{z}_j}\right)\\ &= (g')^{k\bar{l}}g^{i\bar{j}}\frac{\partial^4\phi}{\partial z_i\partial\bar{z}_j\partial z_k\partial\bar{z}_l} + (g')^{k\bar{l}}\frac{\partial^2 g^{i\bar{j}}}{\partial z_k\partial\bar{z}_l}\frac{\partial^2\phi}{\partial z_i\partial\bar{z}_j}.\end{aligned}$$

We substitute $\frac{\partial^4\phi}{\partial z_i\partial\bar z_j\partial z_k\partial\bar z_l}$ in $\Delta'(\Delta\phi)$, so the above reads

$$\Delta'(\Delta\phi) = -g^{k\bar l}(g')^{i\bar j}\frac{\partial^2 g_{i\bar j}}{\partial z_k\partial\bar z_l} + g^{k\bar l}(g')^{t\bar j}(g')^{i\bar s}\frac{\partial^3\phi}{\partial z_t\partial\bar z_s\partial\bar z_l}\frac{\partial^3\phi}{\partial z_i\partial\bar z_j\partial z_k}$$
$$+ g^{k\bar l}g^{i\bar j}\frac{\partial^2 g_{i\bar j}}{\partial z_k\partial\bar z_l} + \Delta F + (g')^{k\bar l}\frac{\partial^2 g^{i\bar j}}{\partial z_k\partial\bar z_l}\frac{\partial^2\phi}{\partial z_i\partial\bar z_j},$$

which we can rewrite after substituting $\frac{\partial^2 g_{i\bar j}}{\partial z_k\partial\bar z_l} = -R_{i\bar j k\bar l}$ and $\frac{\partial^2 g^{i\bar j}}{\partial z_k\partial\bar z_l} = R_{i\bar j k\bar l}$ as

$$\Delta'(\Delta\phi) = \Delta F + g^{k\bar l}(g')^{t\bar j}(g')^{i\bar s}\phi_{t\bar s l}\phi_{i\bar j k} + (g')^{i\bar j}R_{i\bar j k\bar l} - g^{i\bar j}R_{i\bar j k\bar l} + (g')^{k\bar l}R_{i\bar j k\bar l}\phi_{i\bar j}.$$

Let us now rechoose the coordinates so that both g and ϕ are in diagonal form, $g_{i\bar j} = \delta_{ij}$ and $\phi_{i\bar j} = \delta_{ij}\phi_{i\bar i}$, then $(g')^{i\bar s} = \frac{\delta_{is}}{1+\phi_{i\bar i}}$ and the above transforms to

$$\Delta'(\Delta\phi) = \frac{1}{1+\phi_{i\bar i}}\frac{1}{1+\phi_{j\bar j}}\phi_{i\bar j k}\phi_{\bar i j\bar k} + \Delta F + R_{i\bar i k\bar k}\left(-1 + \frac{1}{1+\phi_{i\bar i}} + \frac{\phi_{i\bar i}}{1+\phi_{k\bar k}}\right).$$

Set now $C = \inf_{i\neq k} R_{i\bar i k\bar k}$ and observe that

$$\begin{aligned} R_{i\bar i k\bar k}\left(-1 + \frac{1}{1+\phi_{i\bar i}} + \frac{\phi_{i\bar i}}{1+\phi_{k\bar k}}\right) &= \frac{1}{2}R_{i\bar i k\bar k}\frac{(\phi_{k\bar k}-\phi_{i\bar i})^2}{(1+\phi_{i\bar i})(1+\phi_{k\bar k})} \\ &\geq \frac{C}{2}\frac{(1+\phi_{k\bar k}-1-\phi_{i\bar i})^2}{(1+\phi_{i\bar i})(1+\phi_{k\bar k})} \\ &= C\left(\frac{1+\phi_{i\bar i}}{1+\phi_{k\bar k}} - 1\right), \end{aligned}$$

this yields

$$\Delta'(\Delta\phi) \geq \frac{1}{(1+\phi_{i\bar i})(1+\phi_{j\bar j})}\phi_{i\bar j k}\phi_{\bar i j\bar k} + \Delta F + C\left((n+\Delta\phi)\sum_i \frac{1}{1+\phi_{i\bar i}} - n^2\right).$$

We need to apply one more trick to obtain the requested estimates. Namely,

$$\begin{aligned} \Delta'(e^{-\lambda\phi}(n+\Delta\phi)) &= e^{-\lambda\phi}\Delta'(\Delta\phi) + 2\nabla' e^{-\lambda\phi}\nabla'(n+\Delta\phi) + \Delta'(e^{-\lambda\phi})(n+\Delta\phi) \\ &= e^{-\lambda\phi}\Delta'(\Delta\phi) - \lambda e^{-\lambda\phi}(g')^{i\bar i}\phi_i(\Delta\phi)_{\bar i} - \lambda e^{-\lambda\phi}(g')^{i\bar i}\phi_{\bar i}(\Delta\phi)_i \\ &\quad - \lambda e^{-\lambda\phi}\Delta'\phi(n+\Delta\phi) + \lambda^2 e^{-\lambda\phi}(g')^{i\bar i}\phi_i\phi_{\bar i}(n+\Delta\phi) \\ &\geq e^{-\lambda\phi}\Delta'(\Delta\phi) - e^{-\lambda\phi}(g')^{i\bar i}(n+\Delta\phi)^{-1}(\Delta\phi)_i(\Delta\phi)_{\bar i} \\ &\quad - \lambda e^{-\lambda\phi}\Delta'\phi(n+\Delta\phi), \end{aligned}$$

which follows from the Schwarz Lemma applied to the middle two terms. We will write out one term here, the other goes in an analagous way,

$$\begin{aligned} &(\lambda e^{-\frac{\lambda}{2}\phi}\phi_i(n+\Delta\phi)^{\frac{1}{2}})(e^{-\frac{\lambda}{2}\phi}(\Delta\phi)_{\bar i}(n+\Delta\phi)^{-\frac{1}{2}}) \\ \leq &\frac{1}{2}(\lambda^2 e^{-\bar\lambda\phi}\phi_i\phi_{\bar i}(n+\Delta\phi) + e^{-\lambda\phi}(\Delta\phi)_{\bar i}(\Delta\phi)_i(n+\Delta\phi)^{-1}). \end{aligned}$$

Consider now the following

$$\begin{aligned}
&-(n+\Delta\phi)^{-1}\frac{1}{1+\phi_{i\bar{i}}}(\Delta\phi)_i(\Delta\phi)_{\bar{i}}+\Delta'\Delta\phi \\
\geq&-(n+\Delta\phi)^{-1}\frac{1}{1+\phi_{i\bar{i}}}|\phi_{k\bar{k}i}|^2+\Delta F \\
&+\frac{1}{1+\phi_{i\bar{i}}}\frac{1}{1+\phi_{k\bar{k}}}\phi_{k\bar{i}\bar{j}}\phi_{i\bar{k}j}+C(n+\Delta\phi)\frac{1}{1+\phi_{i\bar{i}}}.
\end{aligned}$$

On the other hand, using the Schwarz inequality, we have

$$\begin{aligned}
&(n+\Delta\phi)^{-1}\frac{1}{1+\phi_{i\bar{i}}}|\phi_{k\bar{k}i}|^2 \\
&=(n+\Delta\phi)^{-1}\frac{1}{1+\phi_{i\bar{i}}}\left|\frac{\phi_{k\bar{k}i}}{(1+\phi_{k\bar{k}})^{\frac{1}{2}}}(1+\phi_{k\bar{k}})^{\frac{1}{2}}\right|^2 \\
&\leq(n+\Delta\phi)^{-1}\left(\frac{1}{1+\phi_{i\bar{i}}}\frac{1}{1+\phi_{k\bar{k}}}\phi_{k\bar{k}i}\phi_{\bar{k}k\bar{i}}\right)\left(1+\phi_{l\bar{l}}\right) \\
&=\frac{1}{1+\phi_{i\bar{i}}}\frac{1}{1+\phi_{k\bar{k}}}\phi_{k\bar{k}i}\phi_{\bar{k}k\bar{i}} \\
&=\frac{1}{1+\phi_{i\bar{i}}}\frac{1}{1+\phi_{k\bar{k}}}\phi_{i\bar{k}k}\phi_{k\bar{i}\bar{k}} \\
&\leq\frac{1}{1+\phi_{i\bar{i}}}\frac{1}{1+\phi_{k\bar{k}}}\phi_{i\bar{k}j}\phi_{k\bar{i}\bar{j}},
\end{aligned}$$

so we get

$$-(n+\Delta\phi)^{-1}\frac{1}{1+\phi_{i\bar{i}}}(\Delta\phi)_i(\Delta\phi)_{\bar{i}}+\Delta'\Delta\phi\geq\Delta F+C(n+\Delta\phi)\frac{1}{1+\phi_{i\bar{i}}}.$$

Putting all these together, we have

$$\Delta'(e^{-\lambda\phi}(n+\Delta\phi)\geq e^{-\lambda\phi}\left(\Delta F+C(n+\Delta\phi)\frac{1}{1+\phi_{i\bar{i}}}\right)-\lambda e^{-\lambda\phi}\Delta'\phi(n+\Delta\phi).$$

Choose λ now to be $-C+1$, then we obtain the following estimate

$$\begin{aligned}
\Delta'(e^{-\lambda\phi}(n+\Delta\phi)\geq&-c_1e^{-\lambda\phi}-c_2e^{-\lambda\phi}(n+\Delta\phi) \\
&+e^{-\lambda\phi}\sum_i\frac{1}{1+\phi_{i\bar{i}}}(n+\Delta\phi),
\end{aligned}$$

where we have used the fact that

$$\Delta'\phi=\sum_i\frac{\phi_{i\bar{i}}}{1+\phi_{i\bar{i}}}=n-\sum_i\frac{1}{1+\phi_{i\bar{i}}}.$$

Now let us notice the following inequality

$$\sum_i \frac{1}{1+\phi_{i\bar{i}}} \geq \left(\frac{\sum_i (1+\phi_{i\bar{i}})}{\prod_i (1+\phi_{i\bar{i}})}\right)^{\frac{1}{n-1}} = e^{-\frac{F}{n-1}} (n+\Delta\phi)^{\frac{1}{n-1}},$$

which can be verified by taking the $(n-1)$-th power of both sides. So the last term in the above can be estimated by

$$e^{-\lambda\phi} \sum_i \frac{1}{1+\phi_{i\bar{i}}} (n+\Delta\phi) \geq e^{-\frac{F}{n-1}} e^{-\frac{\lambda}{n-1}} (e^{-\lambda\phi}(n+\Delta\phi))^{\frac{n}{n-1}}.$$

Setting now $u = e^{-\lambda\phi}(n+\Delta\phi)$ and recalling that $\phi \leq -1$ and hence $e^{-\lambda\phi} \geq 1$,we have finally obtained the following estimate

$$\Delta' u \geq -c_1 - c_2 u + c_0 u^{\frac{n}{n-1}}.$$

Assume that u achieves its maximum at x_0, then at this point, $\Delta' u \leq 0$ and therefore the maximum principle gives us an upper bound $u(x_0) \leq C$ which in its turns gives

$$0 \leq (n+\Delta\phi)(x) \leq e^{\lambda\phi(x)} u(x_0) \leq C$$

and hence we found a C^2-estimate of ϕ.

Step 3 The final step will not be performed here, it is similar to step 2, but the argument will be longer and can be found in the appendix of [27]. This last step will give a bound for $\|\nabla^3\phi\|_{C^0}$. With this bound, we are done. This last bound will namely give us the equicontinuity of ϕ_i and together with the a priori estimate $\|\phi_i - c_i\|_3 \leq C$, the Arzela-Ascoli lemma now tells us that S is compact and in particular that S is closed. □

5.2 Kähler-Einstein metrics for manifolds with $c_1(M) < 0$

In this section, we will first state a theorem proved independently by Aubin and Yau. Its proof is easier than that of the Calabi-Yau theorem in the last section. Then we will give a simple application of Kähler-Einstein metrics. We will end this section with an open question, which is a generalization of the application to the symplectic manifolds.

Theorem 5.8 *(Aubin, Yau) Let M be any compact Kähler manifold with $c_1(M) < 0$, then there exists a unique Kähler-Einstein metric ω with* $\mathrm{Ric}(\omega) = -\omega$.

Remark 5.9 The statement of this theorem is equivalent to the solvability of the following Monge-Ampére equation

$$(\omega_0 + \partial\bar{\partial}\phi)^n = e^{h_{\omega_0}+\phi}\omega_0^n \qquad \text{where} \qquad \partial\bar{\partial}h_{\omega_0} = \mathrm{Ric}(\omega_0) + \omega_0.$$

The plus sign here makes life a lot easier. In fact, we can use the Maximum Principle to get a priori C^0-estimates for solutions of the above equation.

Application (Yau, [28]) There exists a unique complex surface M homotopic to $\mathbb{C}P^2$.

Proof. We already have the following information about the Euler class and the signature, both topological invariants

$$\chi(M) = \int_M c_2(M) = \chi(\mathbb{C}P^2) = 3$$

$$\tau(M) = \frac{1}{3}(c_1(M)^2 - 2c_2(M)) = \tau(\mathbb{C}P^2) = 1,$$

which follows from the Hirzebruch-signature formula and which implies that

$$c_1(M)^2 = c_1(\mathbb{C}P^2)^2 = 3^2 = 9.$$

Since $H^2(\mathbb{C}P^2, \mathbb{Z}) = \mathbb{Z}$, we have that $H^2(M, \mathbb{Z}) = \mathbb{Z}$. Fix a generator $\omega > 0$ of $H^2(M, \mathbb{Z})$ so that we have $c_1(M) = \lambda\omega$ and since $\omega^2 = 1$, we have that $\lambda^2 = 9$ so $\lambda = \pm 3$.

If $\lambda = 3$, then $c_1(M) > 0$ and this implies that there exists a holomorphic $S^2 \subset M$ which implies in its turn that $M = \mathbb{C}P^2$. For this last implication see Chapter 4 of [11].

If $\lambda = -3$, then $c_1(M) < 0$ and there exists a Kähler-Einstein metric ω_{KE}. Now $3\chi(\mathbb{C}P^2) = 3c_2(M) = c_1(M)^2$ and by the uniformization theorem, we deduce that $M = D^2/\Gamma$, so $\pi_1(M) = \Gamma \neq \{\mathrm{id}\}$, which gives us a contradiction, so only $\lambda = 3$ is possible and that proves the application. □

One is now automatically led to the following generalization: *Does there exist a unique symplectic surface M homotopic to $\mathbb{C}P^2$?*

Analogously to the above, we can deal with the case $\lambda = 3$, but it is not clear how to show that $\lambda = -3$ leads to a contradiction. Taubes showed that one can use the Seiberg-Witten theory to deal with the case that M is diffeomorphic to $\mathbb{C}P^2$, but the general case is still open.

Chapter 6

Kähler-Einstein metrics with positive scalar curvature

In this chapter, we will study Kähler-Einstein manifolds of positive scalar curvature.

We will assume that M is a compact manifold and that $c_1(M) > 0$. Because of the latter, we can choose a metric ω with $[\omega] = \pi c_1(M)$. Then

$$\mathrm{Ric}(\omega) = \omega + \partial\bar{\partial} h_\omega,$$

where we can choose $h = h_\omega$ in such a way that

$$\int_M (e^h - 1)\omega^n = 0.$$

Suppose now that we can deform ω within its Kähler class, so that we can find an

$$\omega_\phi = \omega + \partial\bar{\partial}\phi$$

which is Kähler-Einstein, then we have the following

$$\begin{aligned}\omega + \partial\bar{\partial}\phi &= \mathrm{Ric}(\omega + \partial\bar{\partial}\phi) = -\partial\bar{\partial}\log\det(\omega + \partial\bar{\partial}\phi)\\ &= -\partial\bar{\partial}\log\frac{(\omega + \partial\bar{\partial}\phi)^n}{\omega^n} + \mathrm{Ric}(\omega)\\ &= -\partial\bar{\partial}\log\frac{(\omega + \partial\bar{\partial}\phi)^n}{\omega^n} + \omega + \partial\bar{\partial}h.\end{aligned}$$

This leads us to

$$(\omega + \partial\bar{\partial}\phi)^n = e^{h_\omega - \phi}\omega^n. \tag{$*$}$$

Note that $-\log\det(\omega + \partial\bar{\partial}\phi)$ is not globally defined, but by taking a quotient as above, we have a well-defined notion again.

We have seen in Chapter 3 that this equation is in general not solvable and that for example the non-vanishing of the Calabi-Futaki invariants forms an obstruction to the solvability of $(*)$. It is, however, clear that the solvability of $(*)$ is equivalent to the existence of a Kähler-Einstein metric on M.

The goal of this chapter is to provide sufficient and necessary conditions for the existence of Kähler-Einstein metrics on manifolds with positive first Chern class.

6.1 A variational approach

In this section, we will take a variational approach and find a functional for which $(*)$ is the Euler-Lagrange equation.

First let us discuss some interesting functionals.

Definition 6.1 *Let ω be any Kähler metric. Then the generalized energy is given by*

$$J_\omega(\phi) = \frac{1}{V}\sum_{i=0}^{n-1}\frac{i+1}{n+1}\int_M \partial\phi\wedge\bar\partial\phi\wedge\omega^i\wedge\omega_\phi^{n-1-i}$$

where $V = \int_M \omega^n = [\omega]^n([M])$ and $\omega_\phi = \omega + \partial\bar\partial\phi$.

To illustrate this definition, let us compute $J_\omega(\phi)$ for $n = 1, 2$. If $n = 1$,

$$J_\omega(\phi) = \frac{1}{2V}\int_M \partial\phi\wedge\bar\partial\phi = \frac{1}{2V}\int_M |\partial\phi|^2\omega.$$

It follows that $J_\omega(\phi)$ is positive on any non-constant function. In the case that $n = 2$, we have

$$J_\omega(\phi) = \frac{1}{3V}\int_M \partial\phi\wedge\bar\partial\phi\wedge\omega_\phi + \frac{2}{3V}\int_M \partial\phi\wedge\bar\partial\phi\wedge\omega.$$

At any point x of M, we can choose local coordinates $z_1, \dots, z_n$ such that at x, we have $\omega = \sum dz_i\wedge d\bar z_i$. Because ω is positive definite, we can diagonalize ω and $\partial\bar\partial\phi$ simultaneously so that $\phi_{i\bar j} = \phi_{i\bar i}\delta_{ij}$ at x. Now the integrand of the first integral reads locally

$$\begin{aligned}\partial\phi\wedge\bar\partial\phi\wedge\omega_\phi &= (\phi_i dz_i)\wedge(\phi_{\bar i} d\bar z_i)\wedge((1+\phi_{i\bar i})dz_i\wedge d\bar z_i)\\ &= \Big(|\phi_1|^2(1+\phi_{2\bar 2}) + |\phi_2|^2(1+\phi_{1\bar 1})\Big)\frac{\omega^2}{2}.\end{aligned}$$

Again we see that $J_\omega(\phi)$ is positive, if ϕ is not constant and $\omega + \partial\bar\partial\phi$ is positive.

In general, whenever ω_ϕ is positive definite, $J_\omega(\phi) \geq 0$. This can be easily seen from the above definition.

Lemma 6.2 *Let $\{\phi_t\}$ be a smooth family of functions, then*

$$\frac{d}{dt}J_\omega(\phi_t)|_{t=0} = -\frac{1}{V}\int_M \dot\phi(\omega_\phi^n - \omega^n),$$

where $\phi = \phi_0$ and $\dot\phi = \frac{d}{dt}\phi_t|_{t=0}$.

Proof. The proof is just a simple calculation,

$$\begin{aligned}
\frac{d}{dt}J_\omega(\phi_t) &= -\frac{1}{V}\sum_{i=0}^{n-1}\frac{i+1}{n+1}\int_M\Big(2\dot\phi\partial\bar\partial\phi\wedge\omega^i\wedge\omega_\phi^{n-i-1}\\
&\quad+(n-1-i)\phi\partial\bar\partial\phi\wedge\omega^i\wedge\omega_\phi^{n-i-2}\wedge\partial\bar\partial\dot\phi\Big)\\
&= -\frac{1}{V}\sum_{i=0}^{n-1}\frac{i+1}{n+1}\int_M 2\dot\phi\partial\bar\partial\phi\wedge\omega^i\wedge\omega_\phi^{n-i-1}\\
&\quad-\frac{1}{V}\sum_{i=0}^{n-1}\frac{i+1}{n+1}(n-i-1)\int_M\dot\phi\partial\bar\partial\phi\wedge\omega^i\wedge\omega_\phi^{n-i-2}\wedge(\omega+\partial\bar\partial\phi-\omega)\\
&= -\frac{1}{V}\sum_{i=0}^{n-1}\frac{i+1}{n+1}(n-i+1)\int_M\dot\phi\partial\bar\partial\phi\wedge\omega^i\wedge\omega_\phi^{n-1-i}\\
&\quad+\frac{1}{V}\sum_{i=0}^{n-1}\frac{i+1}{n+1}(n-i-1)\int_M\dot\phi\partial\bar\partial\phi\wedge\omega^{i+1}\wedge\omega_\phi^{n-i-2}\\
&= -\frac{1}{V}\sum_{i=0}^{n-1}\frac{i+1}{n+1}(n-i+1)\int_M\dot\phi\partial\bar\partial\phi\wedge\omega^i\wedge\omega_\phi^{n-1-i}\\
&\quad+\frac{1}{V}\sum_{i=1}^{n-1}\frac{i}{n+1}(n-i)\int_M\dot\phi\partial\bar\partial\phi\wedge\omega^i\wedge\omega_\phi^{n-i-1}\\
&= -\frac{1}{V}\int_M\dot\phi\partial\bar\partial\phi\wedge\omega_\phi^{n-1}-\frac{1}{V}\sum_{i=1}^{n-1}\int_M\dot\phi\partial\bar\partial\phi\wedge\omega^i\wedge\omega_\phi^{n-i-1}\\
&= -\frac{1}{V}\sum_{i=0}^{n-1}\int_M\dot\phi\partial\bar\partial\phi\wedge\omega^i\wedge\omega_\phi^{n-i-1}\\
&= -\frac{1}{V}\sum_{i=0}^{n-1}\int_M\dot\phi(\partial\bar\partial\phi+\omega-\omega)\wedge\omega^i\wedge\omega_\phi^{n-i-1}\\
&= -\frac{1}{V}\sum_{i=0}^{n-1}\int_M\dot\phi\omega^i\wedge\omega_\phi^{n-i}+\frac{1}{V}\sum_{i=0}^{n-1}\int_M\dot\phi\omega^{i+1}\wedge\omega_\phi^{n-i-1}\\
&= \frac{1}{V}\int_M\dot\phi(\omega^n-\omega_\phi^n).
\end{aligned}$$

□

If we define

$$F_\omega^0(\phi)=J_\omega(\phi)-\frac{1}{V}\int_M\phi\wedge\omega^n,\qquad\text{then}\qquad\frac{d}{dt}F_\omega^0(\phi_t)|_{t=0}=-\frac{1}{V}\int_M\dot\phi\omega_\phi^n.$$

This functional has many nice properties. Let us list some of them.

Let us choose another metric, say $\omega' = \omega + \partial\bar{\partial}\psi$ in the same Kähler class. In that case, we have

$$\omega_\phi = \omega + \partial\bar{\partial}\phi = \omega' + \partial\bar{\partial}(\phi - \psi) = \omega'_{\phi-\psi}.$$

We will now see how $F^0_\omega(\phi)$ depends on the choice of representatives of $[\omega]$.

Claim 1 For ω and ω' as above, we have

$$F^0_\omega(\phi) - F^0_{\omega'}(\phi - \psi) = F^0_\omega(\psi)$$

and if we set $F^0_\omega(\phi) = F^0(\omega, \omega_\phi)$, this yields the following cocycle relation

$$F^0(\omega, \omega_\phi) - F^0(\omega_\psi, \omega_\phi) = F^0(\omega, \omega_\psi).$$

Proof. Consider now the following difference

$$F^0_\omega(\phi) - F^0_{\omega'}(\phi - \psi),$$

where ψ is fixed. The above lemma gives us the following

$$\frac{d}{dt}(F^0_\omega(\phi_t) - F^0_{\omega'}(\phi_t - \psi)) = 0$$

for any family $\{\phi_t\}$. Therefore the difference is constant, in particular, choosing $\phi = \psi$, we see that this constant equals $F^0_\omega(\psi)$ and the claim is proved. □

Let $P(M, \omega)$ be the following space of functions

$$\{\phi \in C^\infty(M); \omega + \partial\bar{\partial}\phi > 0\}.$$

Claim 2 The functional F^0_ω is convex on $P(M, \omega)$, this means that for any ϕ_1, ϕ_2 in $P(M, \omega)$,

$$F^0_\omega(\frac{1}{2}(\phi_1 + \phi_2)) \leq \frac{1}{2}\left(F^0_\omega(\phi_1) + F^0_\omega(\phi_2)\right).$$

Proof. Put $\omega' = \omega_{\frac{1}{2}(\phi_1+\phi_2)}$. Using Claim 1, we can deduce

$$\begin{aligned}
&F^0_\omega(\frac{1}{2}(\phi_1 + \phi_2)) - \frac{1}{2}\left(F^0_\omega(\phi_1) + F^0_\omega(\phi_2)\right) \\
&= -\frac{1}{2}\left(F^0_{\omega'}(\frac{1}{2}(\phi_1 - \phi_2)) + F^0_{\omega'}(\frac{1}{2}(\phi_2 - \phi_1))\right) \\
&= \frac{1}{4(n+1)}\sum_{i=0}^{n}\frac{1}{V}\int_M (\phi_1 - \phi_2)\omega'^i \wedge (\omega_1^{n-i} - \omega_2^{n-i}) \\
&= -\frac{1}{4(n+1)}\sum_{i=0}^{n}\frac{1}{V}\int_M \partial(\phi_1 - \phi_2) \wedge \bar{\partial}(\phi_1 - \phi_2) \wedge \omega^i \wedge \sum_{j=0}^{n-i-1} \omega_1^j \wedge \omega_2^{n-1-i-j} \\
&\leq 0,
\end{aligned}$$

the claim is proved. □

The next two claims are obvious.

Claim 3 For any constant $\lambda > 0$, $F^0_{\lambda\omega}(\lambda\phi) = \lambda F^0_\omega(\phi)$.

Claim 4 If $\pi : M' \mapsto M$ is a branched covering, then for any $\phi \in P(M,\omega)$, $F^0_{\pi^*\omega}(\pi^*\phi) = F^0_\omega(\phi)$.

The functional, which has $(*)$ as its Euler-Lagrange equation, is given as follows:

$$F_\omega(\phi) = F^0_\omega(\phi) - \log\left(\frac{1}{V}\int_M e^{h_\omega - \phi}\omega^n\right).$$

Remark 6.3 If $n = 1$, $(*)$ reads $1 + \Delta\phi = e^{h_\omega - \phi}$. By the classical uniformization theorem for Riemann surfaces, $M = S^2$, so we get

$$F_\omega(\phi) = \frac{1}{2V}\int_{S^2} |\partial\phi|^2 - \frac{1}{V}\int_{S^2} \phi\omega - \log\frac{1}{V}\int_{S^2} e^{h_\omega - \phi}\omega.$$

This is exactly the functional in L. Nirenberg's problem of prescribing the Gauss curvature equation on S^2.

We also have the cocycle condition for F_ω: for $\omega' = \omega + \partial\bar\partial\psi$, we have

$$F_\omega(\phi) - F_{\omega'}(\phi - \psi) = F_\omega(\psi).$$

To see this, we first observe that

$$F_\omega(\phi) = F_\omega(\phi + C)$$

for any constant C. So we can always normalize ϕ such that

$$\frac{1}{V}\int_M e^{h_\omega - \phi}\omega^n = 1.$$

Similarly we can choose ψ such that

$$\frac{1}{V}\int_M e^{h_\omega - \psi}\omega^n = 1.$$

Then

$$F_\omega(\phi) = F^0_\omega(\phi)$$

and

$$F_\omega(\psi) = F^0_\omega(\psi).$$

Because

$$\mathrm{Ric}(\omega') = -\partial\bar\partial\log(\omega')^n = \omega' + \partial\bar\partial h_{\omega'},$$
$$\mathrm{Ric}(\omega) = -\partial\bar\partial\log\omega^n = \omega + \partial\bar\partial h_\omega,$$

we get after subtracting the bottom from the top equation that

$$-\partial\bar{\partial}\log\left(\frac{(\omega')^n}{\omega^n}\right) = \partial\bar{\partial}(\psi + h_{\omega'} - h_\omega),$$

which then implies that

$$\frac{1}{V}\int_M e^{h_{\omega'}-(\phi-\psi)}(\omega')^n = \frac{1}{V}\int_M e^{h_\omega-\phi}\omega^n = 1.$$

Then

$$F_{\omega'}(\phi-\psi) = F^0_{\omega'}(\phi-\psi),$$

so the cocycle condition for F follows from Claim 1.

Remark 6.4 The functional F_ω is only the difference of two convex functionals, while in the case of $c_1(M) \leq 0$, the corresponding functional is the sum of F^0_ω and another convex functional, so it is always convex.

6.2 Existence of Kähler-Einstein metrics

In this section, we will prove an analytic criterion for the existence of Kähler-Einstein metrics on Kähler manifolds with positive first Chern class.

Definition 6.5 *We say F_ω is bounded from below if there exists $c = c(\omega) > 0$ such that*

$$F_\omega(\phi) \geq -c.$$

F_ω is proper on $P(M,\omega)$ if there exists an increasing function

$$\mu : \mathbb{R} \to [c(\omega), \infty)$$

satisfying $\lim_{t\to\infty}\mu(t) = \infty$, such that for any $\phi \in P(M,\omega)$,

$$F_\omega(\phi) \geq \mu(J_\omega(\phi)).$$

Remark 6.6 It follows from the cocycle condition of F that the properness of F_ω is independent of ω. More precisely, if ω' is another metric, then $F_\omega(\phi) \geq \mu(J_\omega(\phi))$ implies $F_{\omega'}(\phi') \geq \mu'(J_{\omega'}(\phi')$, where

$$\omega' = \omega + \partial\bar{\partial}\psi,\ \phi' = \phi - \psi$$

and $\mu' = \mu - c$ for some constant c depending only on ω and ω'.

Theorem 6.7 *Assume that M has no non-trivial holomorphic vector fields, then M has a Kähler-Einstein metric if and only if F_ω is proper on $P(M,\omega)$.*

We also have the following generalization of the above theorem. We denote by $\mathrm{Aut}(M)$ the automorphism group of M.

Let G be any maximal compact subgroup of $\mathrm{Aut}(M)$. If ω is a G-invariant Kähler metric, then we define

$$P_G(M,\omega) = \{\phi \in C^\infty(M)\,;\, \omega + \partial\bar{\partial}\phi > 0,\ \phi \text{ is G-invariant }\}.$$

Definition 6.8 *Let ω be G-invariant. Then F_ω is proper on $P_G(M,\omega)$ if there exists an increasing function $\mu : \mathbb{R} \to [c(\omega),\infty)$ satisfying*

$$\lim_{t\to\infty} \mu(t) = \infty,$$

such that for any $\phi \in P_G(M,\omega)$,

$$F_\omega(\phi) \geq \mu(J_\omega(\phi)).$$

Remark 6.9 It follows from the cocycle condition of F that the properness on $P_G(M,\omega)$ is independent of ω.

Theorem 6.10 *M has a Kähler-Einstein metric if and only if F_ω is proper on $P_G(M,\omega)$.*

We will prove Theorem 6.7.

There are two directions to be proved and both will be done using the continuity method. First we assume that F_ω is proper. Consider

$$(\omega + \partial\bar{\partial}\phi)^n = e^{h_\omega - t\phi}\omega^n \qquad (*_t)$$

for $\phi \in P(M,\omega)$ and $0 \leq t \leq 1$. We want to show that $(*_1)$ is solvable. Now by Yau's solution of the Calabi conjecture, we know that $(*_0)$ is solvable and this shows that

$$E = \{t \in [0,1]; (*_s) \text{ is solvable for } s \leq t\}$$

is non-empty. We claim that E is open. To see this, let ϕ_t be a solution of $(*_t)$ and let $|t-t'| \ll 1$. Set

$$\phi = \phi_t + \psi,\ \omega_t = \omega + \partial\bar{\partial}\phi_t,$$

then

$$\begin{aligned}(\omega + \partial\bar{\partial}\phi)^n &= (\omega_t + \partial\bar{\partial}\psi)^n\\ &= e^{h_\omega - t'(\phi_t+\psi)}\omega^n\\ &= e^{-(t'-t)\phi_t - t'\psi}\omega_t^n,\end{aligned}$$

where we have used that ϕ_t is a solution of $(*_t)$ and we assume that ψ is small. Define now the operator

$$\Phi : C^{2,\frac{1}{2}}(M,\mathbb{R}) \to C^{0,\frac{1}{2}}(M,\mathbb{R})$$

by

$$\Phi = \log\left(\frac{(\omega + \partial\bar{\partial}\phi)^n}{\omega_t^n}\right) - (t' - t)\phi_t - t'\psi.$$

Linearizing this operator at $\psi = 0$ and $t' = t$, where we will write Δ_{ω_t} as Δ_t, we get

$$D_\psi\Phi(v) = (\Delta_t + t)(v).$$

As in the proof of the Calabi-Yau Theorem, we would like to show that this is invertible, so that we can use the Implicit Function Theorem. Unfortunately, the extra t-term might destroy the invertibility. Note that

$$\Delta_t v = g_t^{i\bar{j}} \frac{\partial^2 v}{\partial z_i \partial \bar{z}_j}.$$

We will need some lemmas to show that the invertibility of the Laplacian is preserved.

Lemma 6.11 *We have* $\mathrm{Ric}(\omega_t) \geq t\omega_t$ *and the equality holds if and only if* $t = 1$.

Proof. The proof of this lemma follows from the following calculation

$$\begin{aligned}
\mathrm{Ric}(\omega_t) &= -\partial\bar{\partial}\log\omega_t^n \\
&= -\partial\bar{\partial}\log\frac{\omega_t^n}{\omega^n} + \mathrm{Ric}(\omega) \\
&= -\partial\bar{\partial}(h_\omega - t\phi_t) + \omega + \partial\bar{\partial}h_\omega \\
&= \omega + t\partial\bar{\partial}\phi_t \\
&= \omega + t(\omega_t - \omega) \\
&= (1-t)\omega + t\omega_t \\
&\geq t\omega_t
\end{aligned}$$

because $\omega \geq 0$. It is clear that we have equality if and only if $t = 1$. □

Lemma 6.12 *The first eigenvalue of* Δ_t *satisfies* $\lambda_1(\Delta_t) > t$ *if* $t < 1$.

Proof. Recall that we can characterize the first eigenvalue variationally by

$$\lambda_1(\Delta_t) = \inf_{\int_M \phi\omega_t^n = 0, \phi \neq 0} \frac{\int_M |\partial\phi|^2 \omega_t^n}{\int_M \phi^2 \omega_t^n}$$

and it is well known that there exists an eigenfunction u such that $\Delta_t u = -\lambda_1 u$.

Set $g = g_t$ and consider

$$\int_M g^{i\bar{j}} g^{k\bar{l}} u_{ik} u_{\bar{j}\bar{l}} dV_g,$$

where dV_g is the volume form of g. At a given point, we can assume that $g_{i\bar{j}} = \delta_{ij}$, so that the integrand reads

$$\begin{aligned}
u_{ij}u_{\bar{i}\bar{j}} &= (u_{ij}u_{\bar{i}})_{\bar{j}} - u_{ij\bar{j}}u_{\bar{i}} \\
&= \operatorname{div}(u_{ij}u_{\bar{i}}) - u_{ji\bar{j}}u_{\bar{i}} \\
&= \operatorname{div}(u_{ij}u_{\bar{i}}) - u_{j\bar{j}i}u_{\bar{i}} - R_{\bar{s}ji\bar{j}}u_s u_{\bar{i}} \\
&= \operatorname{div}(u_{ij}u_{\bar{i}} - u_{i\bar{i}}u_{\bar{j}}) + u_{j\bar{j}}u_{i\bar{i}} - R_{i\bar{s}}u_s u_{\bar{i}} \\
&= \operatorname{div}(u_{ij}u_{\bar{i}} - u_{i\bar{i}}u_{\bar{i}}) + (\Delta u)^2 - R_{i\bar{s}}u_s u_{\bar{i}} \\
&= \operatorname{div}(u_{ij}u_{\bar{i}} - u_{i\bar{i}}u_{\bar{i}}) + (\lambda_1 u)^2 - R_{i\bar{s}}u_s u_{\bar{i}}.
\end{aligned}$$

This implies

$$0 \le \int_M \Big((\Delta u)^2 - \operatorname{Ric}(\partial u, \bar{\partial} u) \Big) dV_g,$$

and hence

$$\lambda_1^2 \int_M u^2 dV_g \ge \int_M \operatorname{Ric}(\partial u, \bar{\partial} u) dV_g > t \int_M |\partial u|^2 dV_g = t\lambda_1 \int_M u^2 dV_g,$$

where we have used the previous lemma and the variational characterization of λ_1, so we have that $\lambda_1 > t$. □

Remark 6.13 If $t = 1$, then $\operatorname{Ric}(\omega_1) = \omega_1$ and $\lambda_1 \ge 1$. The equality holds if and only if

$$\int_M g^{i\bar{j}} g^{k\bar{l}} u_{ik} u_{\bar{j}\bar{l}} dV_g = 0.$$

This is equivalent to saying $u_{\bar{i}\bar{k}} = 0$, which is in its turn equivalent to

$$(g^{i\bar{j}} u_{\bar{j}})_{\bar{k}} = 0,$$

since the metric is parallel. This then gives us a holomorphic vector field

$$X = g^{i\bar{j}} u_{\bar{j}} \frac{\partial}{\partial z_i}.$$

Conversely, if X is a holomorphic vector field satisfying

$$i_X \omega = \bar{\partial} u,$$

then u is an eigenfunction with eigenvalue 1. So we have established the following identification

$$\eta(M) \simeq \{ \text{ eigenfunctions of } \omega_1 \text{ with eigenvalue } 1 \}.$$

It follows from $\lambda_1 > t$ that $D_\psi \Phi$ is invertible for $0 < t < 1$ and by the Implicit Function Theorem, $(*_{t'})$ is solvable for $|t - t'|$ sufficiently small, so we have shown that E is open in $(0,1)$. One can also prove that E is open near $t = 0$ by modifying the above arguments slightly.

To prove that E is closed, we will proceed as before and we will try to find an a priori estimate $\|\phi_t\|_{C^{2,\frac{1}{2}}} \le C$ for every solution $\phi_t \in P(M,\omega)$ of

$$(\omega + \partial\bar{\partial}\phi)^n = e^{h_\omega - t\phi}\omega^n, \qquad (*_t)$$

where $C = C(t, M, \omega)$.

Lemma 6.14 *If ϕ is a solution of $(*_t)$, then there exist uniform constants $b, c > 0$, such that*

$$n + \Delta\phi \le ce^{b(\sup_M \phi - \inf_M \phi)}.$$

Remark 6.15 The proof of this lemma uses the arguments in the C^2-estimate in Yau's proof of the Calabi Conjecture. Note that this lemma implies

$$\|\phi\|_{C^2} \le ce^{b(\sup_M \phi - \inf_M \phi)}.$$

Here one needs to use the fact that $\omega + \partial\bar{\partial}\phi > 0$.

Exactly as in the Calabi-Yau Theorem, we will be able to prove the following lemma (the proof will be omitted) and it remains then to find a bound for the C^0-norm.

Lemma 6.16 *There exists a constant $C = C(\|\phi\|_{C^0})$, such that when ϕ is a solution of $(*_t)$, we have*

$$\|\phi\|_{C^{2,\frac{1}{2}}} \le C.$$

Lemma 6.17 *(Sobolev inequality) Let (X^n, ds^2) be any compact Riemannian manifold (of complex dimension n) with $\mathrm{Ric}(ds^2) \ge \varepsilon ds^2$ for some $\varepsilon > 0$ and $\mathrm{Vol}(ds^2) \ge v > 0$, then there exists a constant $\sigma = \sigma(\varepsilon, v)$ such that for all $f \in C^\infty(X)$,*

$$\left(\int_X |f|^{\frac{2n}{n-1}} dV\right)^{\frac{n-1}{n}} \le \sigma\left(\int_X |\nabla f|^2 dV + \int_X |f|^2 dV\right).$$

Remark 6.18 The proof can be found in [17] and is based on a result by C. Croke. Note that this result is stronger than the usual Sobolev inequality because here the constant σ is independent of the metric. This will turn out to be very useful because we will be comparing different metrics.

The following lemma tells us that to bound the C^0-norm, it suffices to find a bound for $J_\omega(\phi)$ and once we have done that, we will have the required a priori estimate to prove closedness and hence the existence of a Kähler-Einstein metric.

Lemma 6.19 *If ϕ is a solution of $(*_t)$, then there exists a constant $C = C(t)$ so that*

$$\|\phi\|_{C^0} \leq C(1 + J_\omega(\phi)).$$

Proof. Let $\omega_\phi = \omega + \partial\bar{\partial}\phi$ and define the functional $I_\omega(\phi)$, which is the difference of two averages of ϕ with respect to two Kähler metrics, and which satisfies the following estimate

$$\begin{aligned} I_\omega(\phi) &= \frac{1}{V}\int_M \phi(\omega^n - \omega_\phi^n) \\ &= \frac{1}{V}\int_M \phi(\omega - \omega_\phi) \wedge (\omega^{n-1} + \cdots + \omega_\phi^{n-1}) \\ &= \frac{1}{V}\int_M \partial\phi \wedge \bar{\partial}\phi \wedge (\omega^{n-1} + \cdots + \omega_\phi^{n-1}) \\ &\leq (n+1)J_\omega(\phi). \end{aligned}$$

So it suffices to prove the estimate in the lemma with $I_\omega(\phi)$ in place of $J_\omega(\phi)$. Use, as before, the Green formula to get

$$\phi(x) = \frac{1}{V}\int_M \phi(y)\omega(y)^n + \frac{1}{V}\int_M -\Delta\phi(y)G(x,y)\omega(y)^n,$$

where G is chosen so that $G \geq 0$. Since $-\Delta\phi \leq n$, this yields

$$\phi(x) \leq \frac{1}{V}\int_M \phi\omega^n + C,$$

which gives us one half of the required estimate, namely,

$$0 \leq \sup_M \phi \leq \frac{1}{V}\int_M \phi\omega^n + C.$$

Claim

$$0 \leq -\inf_M \phi \leq C\Big(\frac{1}{V}\int_M (-\phi)\omega_\phi^n + 1\Big).$$

Proof. Denote by Δ_t the Laplacian of ω_ϕ, where this metric will also sometimes be denoted by ω_t, which should not cause any confusion. Then, because

$$\omega + \partial\bar{\partial}\phi > 0,$$

we see that

$$\omega = \omega_\phi - \partial\bar{\partial}\phi > 0$$

and taking the trace of this latter expression with respect to ω_ϕ, we get

$$n - \Delta_t\phi = \mathrm{tr}_{\omega_\phi}\,\omega > 0.$$

Define now $\phi_-(x) = \max\{-\phi(x), 1\} \geq 1$, so that

$$\phi_-^p(n - \Delta_t\phi) \geq 0.$$

Integrating this, we get

$$\begin{aligned}
0 &\leq \frac{1}{V}\int_M \phi_-^p(n - \Delta_t\phi)\omega_\phi^n \\
&= \frac{n}{V}\int_M \phi_-^p\omega_\phi^n + \frac{1}{V}\int_M \nabla_t\phi_-^p\nabla_t\phi\,\omega_\phi^n. \\
&= \frac{n}{V}\int_M \phi_-^p\omega_\phi^n + \frac{1}{V}\int_{\{\phi\leq -1\}} \nabla_t\phi_-^p\nabla_t\phi\,\omega_\phi^n \\
&= \frac{n}{V}\int_M \phi_-^p\omega_\phi^n + \frac{1}{V}\int_M \nabla_t\phi_-^p\nabla_t(-\phi_-)\omega_\phi^n \\
&= \frac{n}{V}\int_M \phi_-^p\omega_\phi^n - \frac{1}{V}\frac{4p}{(p+1)^2}\int_M |\nabla_t\phi_-^{\frac{p+1}{2}}|^2\omega_\phi^n.
\end{aligned}$$

Using the fact that $\phi_- \geq 1$ (and hence $\phi_-^p \leq \phi_-^{p+1}$), this yields,

$$\frac{1}{V}\int_M |\nabla_t\phi_-^{\frac{p+1}{2}}|^2\omega_\phi^n \leq \frac{n(p+1)^2}{4pV}\int_M \phi_-^{p+1}\omega_\phi^n.$$

Observe that as before, when ϕ is a solution of $(*_t)$, we have

$$\mathrm{Ric}(\omega_t) \geq t\omega_t.$$

Since ω and ω_t are in the same Kähler class, we get

$$\int_M \omega_t^n = \int_M \omega^n = V.$$

Using the Sobolev inequality in Lemma 6.17, we see

$$\frac{1}{V}\left(\int_M |\phi_-|^{\frac{(p+1)n}{n-1}}\omega_t^n\right)^{\frac{n-1}{n}} \leq \frac{c(p+1)}{V}\int_M \phi_-^{p+1}\omega_t^n.$$

Once more, Moser's iteration will show us that

$$\sup_M \phi_- = \lim_{p\to\infty}\|\phi_-\|_{L^{p+1}(M,\omega_t)} \leq C\|\phi_-\|_{L^2(M,\omega_t)}.$$

Recall that $\lambda_1(\omega_t) \geq t$, so that the Poincaré inequality reads

$$\frac{1}{V}\int_M\left(\phi_- - \frac{1}{V}\int_M \phi_-\omega_t^n\right)^2\omega_t^n \leq \frac{1}{tV}\int_M |\nabla\phi_-|^2\omega_t^n$$
$$\leq \frac{C}{tV}\int_M \phi_-\omega_t^n,$$

where we have set $p = 1$ and used the same reasoning as before. This then implies that

$$\max\{-\inf_M \phi, 1\} = \sup_M \phi_- \leq \frac{C(t)}{V}\int_M \phi_-\omega_t^n.$$

Since

$$\int_M e^{-h_\omega + t\phi}\omega_t^n = V,$$

we can easily deduce

$$\frac{1}{V}\int_{\phi>0} \phi\omega_t^n \leq C/t.$$

Combining this together with the above, we get

$$-\inf_M \phi \leq \frac{C}{V}\int_M (-\phi)\omega_t^n + C,$$

which proves the claim.

Since

$$\int_M e^{h_\omega - t\phi}\omega_t^n = V,$$

we have

$$-\frac{1}{V}\int_{\phi<0} \phi\omega^n \leq C/t.$$

Then the lemma follows because

$$\|\phi\|_{C^0} = \sup_M |\phi| = \max\{\sup_M \phi, -\inf_M \phi\} \leq \sup_M \phi - \inf_M \phi,$$

and from the two obtained estimates, we get

$$\|\phi\|_{C^0} \leq C(I_\omega(\phi) + 1) \leq C(J_\omega(\phi) + 1).$$

□

So all that remains to be done is to show that $J_\omega(\phi)$ is bounded. Recall that

$$\frac{d}{dt}J_\omega(\phi_t) = \frac{1}{V}\int_M \dot{\phi}_t(\omega^n - \omega_t^n),$$

$$\frac{d}{dt}I_\omega(\phi_t) = \frac{1}{V}\frac{d}{dt}\int_M \phi_t(\omega^n - \omega_t^n) = \frac{1}{V}\int \dot{\phi}_t(\omega^n - \omega_t^n) - \frac{1}{V}\int_M \phi_t\dot{\omega}_t^n,$$

and since

$$\dot{\omega}_t^n = n\dot{\omega}_t \wedge \omega_t^{n-1} = n\partial\bar{\partial}\dot{\phi}_t \wedge \omega_t^{n-1} = \Delta_t\dot{\phi}_t\omega_t^n,$$

we obtain the following expression

$$\frac{d}{dt}\Big(I_\omega(\phi_t) - J_\omega(\phi_t)\Big) = -\frac{1}{V}\int_M \phi_t\Delta_t\dot{\phi}_t\omega_t^n.$$

Remark 6.20 We know that $\lambda_1(\omega_t) > t$ for $t < 1$. Let $\{\phi_t\}$ be a continuous family of solutions of $(*_t)$, then this family is also smooth in t.

Let ϕ_t be a solution of $(*_t)$, so

$$(\omega + \partial\bar{\partial}\phi_t)^n = e^{h_\omega - t\phi_t}\omega^n.$$

Differentiate this expression with respect to t

$$n\partial\bar{\partial}\dot{\phi}_t \wedge (\omega + \partial\bar{\partial}\phi_t)^{n-1} = (-\phi_t - t\dot{\phi}_t)e^{h_\omega - t\phi_t}\omega^n,$$

this implies that

$$\Delta_t\dot{\phi}_t\omega_t^n = (-\phi_t - t\dot{\phi}_t)\omega_t^n,$$

which is the same as

$$\Delta_t\dot{\phi}_t = (-\phi_t - t\dot{\phi}_t).$$

This means that we can rewrite the above as

$$\begin{aligned}\frac{d}{dt}\Big(I_\omega(\phi_t) - J_\omega(\phi_t)\Big) &= \frac{1}{V}\int_M \phi_t(\phi_t + t\dot{\phi}_t)\omega_t^n \\ &= -\frac{d}{dt}\Big(\int_M \phi_t e^{h_\omega - t\phi_t}\omega^n\Big) + \frac{1}{V}\int_M \dot{\phi}_t e^{h_\omega - t\phi_t}\omega^n.\end{aligned}$$

Since we know that for all t,

$$\int_M e^{h_\omega - t\phi_t}\omega^n = V,$$

differentiating with respect to t yields

$$\int_M (\phi_t + t\dot{\phi}_t)e^{h_\omega - t\phi_t}\omega^n = 0,$$

so that we can simplify the above even further to

$$\frac{d}{dt}\Big(I_\omega(\phi_t) - J_\omega(\phi_t)\Big) = -\frac{d}{dt}\Big(\frac{1}{V}\int_M \phi_t\omega_t^n\Big) - \frac{1}{tV}\int_M \phi_t e^{h_\omega - t\phi_t}\omega^n.$$

Multiplying it by t, we get

$$\frac{d}{dt}\Big(t(I_\omega(\phi_t) - J_\omega(\phi_t))\Big) - (I_\omega(\phi_t) - J_\omega(\phi_t)) = -\frac{d}{dt}\Big(\frac{t}{V}\int_M \phi_t \omega_t^n\Big). \tag{a}$$

Integrating this from 0 to t gives us

$$t(I_\omega(\phi_t) - J_\omega(\phi_t)) - \int_0^t (I_\omega(\phi_s) - J_\omega(\phi_s))ds = -\Big(\frac{t}{V}\int_M \phi_t \omega_t^n\Big),$$

which is equivalent, using the definition on I_ω and J_ω, to

$$-\int_0^t (I_\omega(\phi_s) - J_\omega(\phi_s))ds = t\Big(J_\omega(\phi_t) - \frac{1}{V}\int_M \phi_t \omega^n\Big). \tag{b}$$

An easy manipulation shows that

$$I_\omega(\phi_s) - J_\omega(\phi_s) = \frac{1}{V}\sum_{i=0}^{n-1}\frac{n-i}{n+1}\int_M \partial\phi \wedge \bar\partial\phi \wedge \omega^i \wedge \omega_t^{n-i-1} \geq 0,$$

and therefore,

$$J_\omega(\phi_t) - \frac{1}{V}\int_M \phi_t \omega^n \leq 0.$$

In fact, the function

$$t\Big(J_\omega(\phi_t) - \frac{1}{V}\int_M \phi_t \omega^n\Big)$$

is decreasing in t. We may assume that t is away from 0, since we know that $(*_0)$ is solvable in any case.

However, the above translates into

$$J_\omega(\phi_t) - \frac{1}{V}\int_M \phi_t \omega^n = F_\omega(\phi_t) + \log\Big(\frac{1}{V}\int_M e^{h_\omega - \phi_t}\omega^n\Big).$$

Since

$$\frac{1}{V}\int_M e^{h_\omega - \phi_t}\omega^n = \frac{1}{V}\int_M e^{-(1-t)\phi_t} e^{h_\omega - t\phi_t}\omega^n = \int_M e^{-(1-t)\phi_t}\frac{\omega_t^n}{V}$$

and

$$\int_M \frac{\omega_t^n}{V} = 1,$$

together with the fact that the logarithm is a concave function, we get

$$\begin{aligned}\log\Big(\frac{1}{V}\int_M e^{h_\omega - \phi_t}\omega^n\Big) &= \log\Big(\int_M e^{-(1-t)\phi_t}\frac{\omega_t^n}{V}\Big)\\ &\geq \int_M -(1-t)\phi_t \frac{\omega_t^n}{V}\\ &= -\frac{1-t}{V}\int_M \phi_t \omega_t^n.\end{aligned}$$

This leads to the following estimate

$$\begin{aligned} F_\omega(\phi_t) &\le -\log\left(\frac{1}{V}\int_M e^{h_\omega-\phi_t}\omega^n\right) \\ &\le \frac{1-t}{V}\int_M \phi_t\omega_t^n, \end{aligned}$$

since

$$J_\omega(\phi_t) - \frac{1}{V}\int_M \phi_t\omega^n \le 0.$$

We will now, finally, use the fact the $F_\omega(\phi_t)$ is a proper function, which is the assumption under which we started, and which means

$$F_\omega(\phi_t) \ge \mu(J_\omega(\phi_t))$$

for some increasing function μ. Furthermore, we have

$$\frac{1}{V}\int_M e^{h_\omega}\omega^n = \frac{1}{V}\int_M e^{h_\omega - t\phi_t}\omega^n = 1.$$

This implies that ϕ_t must be negative somewhere and hence $-\inf_M \phi_t$ is positive. So we have

$$0 \le -\inf_M \phi_t \le C\left(\frac{1}{V}\int_M (-\phi_t)\omega_t^n + C'\right).$$

This follows

$$\frac{1}{V}\int_M \phi_t\omega_t^n \le C.$$

Since $t \in (0,1]$, we have,

$$\frac{1-t}{V}\int_M \phi_t\omega_t^n \le C(1-t).$$

This in its turn tells us that $J_\omega(\phi_t)$ is bounded so that we have the desired estimate

$$\|\phi\|_{C^0} \le C(1 + J_\omega(\phi_t)) \le C.$$

This yields

$$\|\phi\|_{C^{2,\frac{1}{2}}} \le C.$$

This final estimate enables us to deduce that the set E is open and closed. Hence, $(*_1)$ is solvable and its solution provides us with a Kähler-Einstein metric.

Now we prove the other part of Theorem 6.7. Assume that there exists a Kähler-Einstein metric. Because the properness is independent of the metric, we can choose our metric ω such that $\mathrm{Ric}(\omega) = \omega$. Thus it suffices to prove the following theorem.

Theorem 6.21 *Let (M, ω) be a Kähler-Einstein manifold with* $\mathrm{Ric}(\omega) = \omega$. *Then there exist constants $\delta = \delta(n)$ and $c = c(n, \lambda_2(\omega) - 1) \geq 0$ such that for any $\phi \in P(M, \omega)$ which satisfy $\phi \perp \Lambda_1$, we have*

$$F_\omega(\phi) \geq J_\omega(\phi)^\delta - c,$$

which is the same as

$$\frac{1}{V} \int_M e^{-\phi} \omega^n \leq C e^{J_\omega(\phi) - \frac{1}{V} \int_M \phi \omega^n - J_\omega(\phi)^\delta}.$$

Here Λ_1 is the space of eigenfunctions with eigenvalue 1 and by $\phi \perp \Lambda_1$, we mean

$$\int_M \phi \psi \omega^n = 0$$

for all $\psi \in \Lambda_1$. Moreover, $\lambda_2(\omega)$ denotes the smallest eigenvalue of ω which is greater than 1. Note that if $\eta(M) = \{0\}$, then $\phi \perp \Lambda_1$ is vacuous and the above inequality holds for any ϕ.

Remark 6.22 This theorem is a stronger statement than just showing properness for some increasing function, because we have the function μ here explicitly.

If $n = 1$, then $M = S^2$ because of positivity of $C_1(M)$. If we further assume instead that ϕ is perpendicular to the eigenfunctions corresponding to the eigenvalue 1, then Aubin showed

$$F_\omega(\phi) \geq \varepsilon J_\omega(\phi) - C.$$

Conjecture 6.23 *(Tian) Under the assumption of Theorem* 6.21, *one can find $\varepsilon = \varepsilon(n)$ and $C = C(n, \lambda_1 - 1)$ for all n such that*

$$F_\omega(\phi) \geq \varepsilon J_\omega(\phi) - C.$$

Before proving Theorem 6.21, we will first explain why we make the assumption on the algebra of holomorphic vector fields. Suppose that $X \in \eta(M)$ for $X \neq 0$. Denote by Φ_t the flow corresponding to $\mathrm{Re}(X)$, then

$$\Phi_t^* \omega = \omega + \partial\bar{\partial}\phi_t.$$

We will write ω_t for $\omega + \partial\bar{\partial}\phi_t$ and as before we can choose ϕ_t such that

$$\frac{1}{V} \int_M e^{h_\omega - \phi_t} \omega^n = 1.$$

Claim

$$\frac{d}{dt} F_\omega(\phi_t) = \mathrm{Re}(f_M(X)),$$

where f_M is the Calabi-Futaki invariant.

Proof. First of all, we have

$$\mathrm{Ric}(\omega) - \omega = \partial\bar{\partial} h_\omega,$$

so

$$\Phi_t^* \,\mathrm{Ric}(\omega) - \Phi_t^*\omega = \partial\bar{\partial}\Phi_t^* h_\omega.$$

Because Φ_t is a diffeomorphism, we also have

$$\frac{1}{V}\int_M e^{\Phi_t^* h_\omega}\Phi_t^*\omega^n = 1,$$

since

$$\frac{1}{V}\int_M e^{h_\omega}\omega^n = 1.$$

Therefore, $h_{\omega_t} = \Phi_t^* h_\omega$, which implies that

$$\dot{h}_{\omega_t} = \mathrm{Re}(X)(h_{\omega_t}).$$

We also have

$$\mathrm{Ric}(\omega_t) - \omega_t = \mathrm{Ric}(\omega) - \omega - \partial\bar{\partial}\log\left(\frac{\omega_t^n}{\omega^n}\right) - \partial\bar{\partial}\phi_t,$$

which in its turn gives

$$h_{\omega_t} = h_\omega - \log\left(\frac{\omega_t^n}{\omega^n}\right) - \phi_t.$$

Differentiating the above with respect to t, we get

$$\dot{h}_{\omega_t} = -\Delta_{\omega_t}\dot{\phi}_t - \dot{\phi}_t,$$

and therefore

$$\begin{aligned}\frac{d}{dt}F_\omega(\phi_t) &= -\frac{1}{V}\int_M \dot{\phi}_t\Phi_t^*\omega^n \\ &= \frac{1}{V}\int_M (\Delta_{\omega_t}\dot{\phi}_t + \mathrm{Re}(X)(h_{\omega_t}))\omega_t^n \\ &= \mathrm{Re}(f_M(X)).\end{aligned}$$

□

The following corollary is an immediate consequence of this claim.

Corollary 6.24 *F_ω is bounded from below only if $f_M = 0$.*

Proof. Suppose that $f_M(X) \neq 0$, then we can assume that $\mathrm{Re}(f_M(X)) < 0$ (since if it was imaginary or positive we could change X to iX or $-X$). Then the claim shows that $F_\omega(\phi_t) = t\,\mathrm{Re}(f_M(X)) \to -\infty$ as $t \to \infty$. □

Proof of Theorem 6.21. For simplicity, we assume that $\eta(M) = \{0\}$. We refer the readers to [24] for the general case.

For any $\phi \in P(M, \omega)$, put $\omega' = \omega_\phi$. Consider

$$(\omega' + \partial\bar{\partial}\psi)^n = e^{h_\omega - t\psi}\omega'^n \qquad (**_t)$$

We may assume that $(**_1)$ has a solution ϕ_1. We will try to use the continuity method backwards, that is starting at ϕ_1. So we set

$$E = \{t \in [0,1]; (**_s) \text{ is solvable for all } s \in [t,1]\}.$$

Clearly, $1 \in E$, so E is non-empty. Showing that E is open goes as before. We use that $\lambda_1 > 1$ (since $\lambda_1 = 1$ would imply that $\eta(M)$ was non-trivial) and that ω is a Kähler-Einstein metric. Then we can use the Implicit Function Theorem at $t = 1$.

To show that E is closed comes down to the following a priori estimate $\|\psi\|_{C^{2,\frac{1}{2}}} \leq C$. Note that C always denotes a constant independent of t.

It was shown in [25] that there are $\delta > 0$, which depends only on M, and C', which may depend on ω', such that for any $\psi \in P(M, \omega')$,

$$\frac{1}{V}\int_M e^{-\delta(\psi - \sup_M \psi)}\omega'^n \leq C'.$$

It follows that if ϕ_t is a solution of $(**_t)$ for $t < \delta/2n$, we have that

$$\sup_M \phi_t \leq C$$

and

$$\frac{1}{V}\int_M \left|e^{h_{\omega'} - t\phi_t} - 1\right|^n \omega'^n \leq C.$$

Put $\psi_+ = \max\{-\phi_t, 1\}$, then for any $p > 0$, we have

$$\begin{aligned}
&\frac{1}{V}\int_M |\partial\psi_+^{\frac{p+1}{2}}|^2\omega'^n \\
=&\frac{n}{V}\int_M \partial\psi_+^{\frac{p+1}{2}} \wedge \bar{\partial}\psi_+^{\frac{p+1}{2}} \wedge \omega'^{n-1} \\
\leq&\frac{n}{V}\int_M \partial\psi_+^{\frac{p+1}{2}} \wedge \bar{\partial}\psi_+^{\frac{p+1}{2}} \wedge \sum_{i=0}^{n-1}\omega'^{n-1-i} \wedge (\omega' + \partial\bar{\partial}\phi_t)^i \\
=&-\frac{n(p+1)^2}{4pV}\int_M \psi_+^p\partial\bar{\partial}\psi_+ \wedge \sum_{i=0}^{n-1}\omega'^{n-1-i} \wedge (\omega' + \partial\bar{\partial}\phi_t)^i \\
=&\frac{n(p+1)^2}{4pV}\int_M \psi_+^p((\omega' + \partial\bar{\partial}\phi_t)^n - \omega'^n)
\end{aligned}$$

$$
\begin{aligned}
&= \frac{n(p+1)^2}{4pV}\int_M \psi_+^p (e^{h_{\omega'}-t\phi_t}-1)\omega'^n\\
&\leq \frac{n(p+1)^2}{4pV}\int_M \psi_+^{p+1}|e^{h_{\omega'}-t\phi_t}-1|\omega'^n\\
&\leq \frac{n(p+1)^2}{4p}\left(\frac{1}{V}\int_M \psi_+^{\frac{2n(p+1)}{2n-1}}\omega'^n\right)^{\frac{2n-1}{2n}}\left(\frac{1}{V}\int_M |e^{h_{\omega'}-t\phi_t}-1|^n\omega'^n\right)^{\frac{1}{2n}}.
\end{aligned}
$$

Using the Sobolev inequality, we can deduce from the above inequalities that

$$
\left(\frac{1}{V}\int_M \psi_+^{\frac{n(p+1)}{n-1}}\omega'^n\right)^{\frac{n-1}{n}} \leq C\left(\frac{1}{V}\int_M \psi_+^{\frac{2n(p+1)}{2n-1}}\omega'^n\right)^{\frac{2n-1}{2n}}.
$$

Then the same arguments as in the proof of Theorem 5.1 yield $\|\phi_t\|_{C^{2,\frac{1}{2}}} \leq C$ for $t < \delta/2n$.

Now we assume that $t \geq \delta/2n$. Then by Lemma 6.19, we have

$$
\begin{aligned}
\|\phi\|_{C^{2,\frac{1}{2}}} &\leq C(1+J_{\omega'}(\phi_t))\\
&\leq C(1+n(I_{\omega'}(\phi_t)-J_{\omega'}(\phi_t))).
\end{aligned}
$$

Lemma 6.25 *Let $\{\phi_t\}$ be a smooth family of solutions of $(**_t)$. Then the function $I_{\omega'}(\phi_t)-J_{\omega'}(\phi_t)$ is monotonically increasing in t.*

Proof. As before, we have

$$
\begin{aligned}
\frac{d}{dt}\Big(I_\omega(\phi_t)-J_\omega(\phi_t)\Big) &= -\frac{1}{V}\int_M \phi_t\Delta_t\dot{\phi}_t\omega_t^n\\
&= -\frac{1}{V}\int_M (\Delta_t\dot{\phi}_t+t\dot{\phi}_t)\Delta_t\dot{\phi}_t\omega_t^n,
\end{aligned}
$$

where $\omega_t = \omega' + \partial\bar{\partial}\phi_t$. We can write $\dot{\phi}_t = \sum_{i=0}^{\infty} c_i\psi_i$, where ψ_i are the eigenfunctions of Δ_t and $\psi_0 = 1$ is the constant eigenfunction. So the above integrand then reads

$$
\sum_{i=1}^{\infty}|c_i|^2(\lambda_i - t)\lambda_i,
$$

which is non-negative since $\lambda_i \geq t$. □

So we end up with

$$
\|\phi\|_{C^{2,\frac{1}{2}}} \leq C(1+n(I_\omega(\phi_t)-J_\omega(\phi_t))) \leq C.
$$

Therefore, we have shown that $(**_t)$ is solvable for all $t \in [0,1]$ and here we have $\phi_1 = -\phi$.

So we have found a smooth family of solutions $\{\phi_t\}$ defining a path in $P(M, \omega)$. This path clearly depends on the base point ϕ. From Lemma 6.25, we know that $I_{\omega'}(\phi_t) - J_{\omega'}(\phi_t)$ is increasing with t.

Using the cocycle relation for F_ω after Remark 6.3, when we set $\phi = 0$, $\psi = \phi$, $\omega' = \omega_\phi$, we get

$$\begin{aligned} F_\omega(\phi) &= -F_{\omega_\phi}(-\phi) \\ &= -F_{\omega'}(\phi_1) \\ &= -J_{\omega'}(\phi_1) + \frac{1}{V}\int_M \phi_1 \omega'^n + \log\left(\frac{1}{V}\int_M e^{h_{\omega'} - \phi_1}\omega'^n\right), \end{aligned}$$

but because of $(**_1)$, we know that

$$\frac{1}{V}\int_M e^{h_{\omega'} - \phi_1}\omega'^n = \frac{1}{V}\int_M \omega^n = 1,$$

so this disappears with the logarithm and therefore it follows from equation (b) that

$$F_\omega(\phi) = \int_0^1 (I_{\omega'}(\phi_t) - J_{\omega'}(\phi_t))dt \geq 0,$$

where we used that the integrand is positive. So we can now state the following corollary.

Corollary 6.26 *(Ding & Tian) $F_\omega(\phi) \geq 0$ for all ϕ with $\omega + \partial\bar{\partial}\phi > 0$.*

This implies in particular the Moser-Trudinger-Onofri inequality on S^2, which reads

$$\frac{1}{4\pi}\int_{S^2} e^{-\phi}\omega \leq e^{\frac{1}{8\pi}\int_{S^2}|\nabla\phi|^2\omega - \frac{1}{4\pi}\int_{S^2}\phi\omega},$$

where we have used that $M = S^2$ and hence $V = 4\pi$. Here, however, we are after the extra $J_\omega(\phi)^\delta$-term in the estimate.

For $t > 0$, we have that

$$\begin{aligned} F_\omega(\phi) &\geq \int_t^1 (I_{\omega'}(\phi_s) - J_{\omega'}(\phi_s))ds \\ &\geq (1-t)(I_{\omega'}(\phi_t) - J_{\omega'}(\phi_t)). \end{aligned}$$

Now we want to show that $J_{\omega'}(\phi_t) \sim J_{\omega'}(\phi_1)$ uniformly for t close to 1, where $\sim$ means that they are comparable, which is at first sight a difficult requirement since ϕ_t and ϕ_1 are solutions of different equations. But if we manage to do this, we will have the following estimate

$$F_\omega(\phi) \geq (1-t)J_{\omega'}(\phi_1) - C.$$

The following lemma will be crucial but will not be proved here, see [24] for a proof.

Lemma 6.27 *(Smoothing lemma) Let $\tilde{\omega}$ be any Kähler metric representing $c_1(M) > 0$ with $\mathrm{Ric}(\tilde{\omega}) \geq (1-\varepsilon)\tilde{\omega}$, where*

$$\mathrm{Ric}(\tilde{\omega}) - \tilde{\omega} = \partial\bar{\partial}h_{\tilde{\omega}}$$

and

$$\frac{1}{V}\int_M (e^{h_{\tilde{\omega}}} - 1)\tilde{\omega}^n = 1.$$

Then there exists

$$\tilde{\omega}' = \tilde{\omega} + \partial\bar{\partial}\phi$$

such that

i) $\|\phi\|_{C^0} \leq 3\|h_{\tilde{\omega}}\|_{C^0}$,

ii) $\|h_{\tilde{\omega}'}\|_{C^{\frac{1}{2}}(\tilde{\omega}')} \leq C(1 + \|h_{\tilde{\omega}}\|^2_{C^0})\varepsilon^{\beta}$,

where $\beta = \beta(n)$ and $C = C(n, \lambda_1(\tilde{\omega}'), \sigma_{\tilde{\omega}'})$.

Remark 6.28 By $C^{\frac{1}{2}}$, we mean the Hölder norm which depends on the metric.

The proof uses Hamilton's Ricci flow

$$\frac{\partial u}{\partial t} = \log\frac{(\tilde{\omega} + \partial\bar{\partial}u)^n}{\tilde{\omega}^n} - u + h_{\tilde{\omega}}$$

with boundary condition $u|_{t=0} = 0$. Setting $\phi = u_1$ and $\tilde{\omega}' = \tilde{\omega} + \partial\bar{\partial}u_1$, this proves the lemma.

Remark 6.29 Notice that although $\|h_{\tilde{\omega}}\|$ might not be small, this lemma ensures that $\|h_{\tilde{\omega}'}\|$ is small as long as ϵ is sufficiently small. Also $\tilde{\omega}'$ is a continuous deformation of $\tilde{\omega}$.

Because of the Kähler-Einstein property and the fact that we have a solution of $(**_t)$, we get that

$$\begin{aligned}\partial\bar{\partial}h_{\omega_t} &= \mathrm{Ric}(\omega_t) - \omega_t\\ &= -\partial\bar{\partial}\log(\omega' + \partial\bar{\partial}\phi_t)^n - \omega' - \partial\bar{\partial}\phi_t\\ &= -\partial\bar{\partial}\log e^{h_{\omega'} - t\phi_t} - \partial\bar{\partial}\log(\omega')^n - \omega' - \partial\bar{\partial}\phi_t\\ &= -\partial\bar{\partial}(1-t)\phi_t,\end{aligned}$$

so we can write

$$h_{\omega_t} = -(1-t)\phi_t + c_t,$$

where c_t is determined by

$$\int_M \left(e^{-(1-t)\phi_t + c_t} - 1\right)\omega_t^n = 0.$$

This immediately implies the following bound on constants c_t,

$$|c_t| \leq (1-t)\|\phi_t\|_{C^0}.$$

Setting $\varepsilon = 1-t$, we get

$$\mathrm{Ric}(\omega_t) \geq (1-\varepsilon)\omega_t = t\omega_t,$$

and

$$||h_{\omega_t}||_{C^0} \leq 2\varepsilon||\phi_t||_{C^0}.$$

Therefore, the above lemma provides us with a $\omega'_t = \omega_t + \partial\bar{\partial}u_t$ which satisfies

$$\begin{array}{ll} 1) & \|u_t\|_{C^0} \leq 3(1-t)\|\phi_t\|_{C^0}, \\ 2) & \|h_{\omega'}\|_{C^{0,\frac{1}{2}}(\omega'_t)} \leq C(1+(1-t)^2\|\phi_t\|^2_{C^0})(1-t)^\beta, \end{array}$$

where $C = C(n, \lambda_1(\omega'_t), \sigma_{\omega_t})$.

We now make the following observation:

$$\omega'_1 = \omega' + \partial\bar{\partial}\phi_1 = \omega + \partial\bar{\partial}\phi - \partial\bar{\partial}\phi = \omega.$$

So in this case, we do not need to deform the metric, because at $t = 1$, we already have a Kähler-Einstein metric. Since ω is a Kähler-Einstein metric, it now follows that there exists a unique ψ_t with $\omega = \omega'_t + \partial\bar{\partial}\psi_t$ and

$$\omega^n = e^{h_{\omega'_t} - \psi_t}\omega'^n_t. \qquad (***_t)$$

Similarly as above, but in addition using the Maximum Principle, we get

$$\phi_t = \phi_1 - \psi_t - u_t + c_t + \mu_t,$$

where μ_t is determined by

$$\int_M \left(e^{h_{\omega_t} - u_t + \mu_t} - 1\right)\omega_t^n = 1.$$

It is easy to see

$$|\mu_t| \leq 10(1-t)\|\phi_t\|_{C^0},$$

and hence, we have that

$$-u_t + c_t + \mu_t \sim (1-t)\|\phi_t\|_{C^0}.$$

Our goal is to control the behavior of $\phi_t - \phi_1$ so that it remains to get ψ_t under control.

The solvability of $(**_t)$ and the Kähler-Einstein property together imply that

$$\omega^n = e^{h_{\omega_t} - (\phi_1 - \phi_t)} \omega_t^n,$$

which unfortunately does not give us any control over $h_{\dot{\omega_t}}$. This makes us study $(***_t)$ instead.

Note that $\psi_1 = 0$ and as before, we will try to use the Implicit Function Theorem here. Consider the following operator

$$\Phi_t : C^{2,\frac{1}{2}}(M, \mathbb{R}) \to C^{0,\frac{1}{2}}(M, \mathbb{R})$$

defined by

$$\psi \mapsto \log \frac{(\omega - \partial\bar{\partial}\psi)^n}{\omega^n} - h_{\omega_t'} - \psi.$$

We would like to show that $D\Phi_t$ is invertible at $\psi = 0$, so that we can deduce that $\Phi_t(\psi) = 0$ is solvable, which is equivalent to saying that $-\psi$ solves $(***_t)$. In order to arrive at this conclusion, we will make the following two observations.

– We know that $\eta(M) = \{0\}$ and that this means that there are no eigenfunctions corresponding to the eigenvalue 1. So $\lambda_1(\omega) > 1$ which shows that the linearization of Φ_t at $\psi = 0$, given by $D\Phi_t|_{\psi=0}(v) = -\Delta_\omega v - v$, is invertible. Therefore, the Implicit Function Theorem tells us that there exists a unique ψ, such that $\Phi_t(\psi) = 0$ and $\|\psi\|_{C^{2,\frac{1}{2}}} \leq C\varepsilon$ if $\|h_{\omega_t'}\|_{C^{0,\frac{1}{2}}} \leq \varepsilon$ for some small ε.

– Suppose we are given two metrics which are comparable, that is $\frac{1}{2}\omega \leq \tilde{\omega} \leq 2\omega$, then $\lambda_1(\tilde{\omega}) \geq 2^{-n-1}\lambda_1(\omega)$ and $\sigma_{\tilde{\omega}} \geq 2^{n+1}\sigma_\omega$. So the corresponding Poincaré and Sobolev constants are also comparable.

Take now some $C \geq C(n, 2^{-n-1}\lambda_1(\omega), 2^{n+1}\sigma_\omega)$ and assume that $C\varepsilon < \frac{1}{4}$, then we can choose a t_0, which may still depend on ϕ, such that

$$\begin{aligned} &(1-t_0)^\beta(1+(1-t_0)^2\|\phi_{t_0}\|_{C^0}^2)^{n+1} \\ =& \sup_{t_0 \leq t \leq 1} (1-t)^\beta(1+(1-t)^2\|\phi_t\|_{C^0}^2)^{n+1} \\ =& \frac{\varepsilon}{4C+1}. \end{aligned}$$

Claim For any $t \in [t_0, 1]$, we have $\|\psi_t\|_{C^{2,\frac{1}{2}}} < \frac{1}{4}$.

Proof. Clearly, $\psi_1 = 0$ satisfies this. If there exists ψ_{t_1} such that $t_1 \in [t_0, 1]$ with $\|\psi_{t_1}\|_{C^{2,\frac{1}{2}}} = \frac{1}{4}$, then because of Remark 6.29, we get

$$\|h_{\omega_t'}\|_{C^{2,\frac{1}{2}}(\omega)} \leq \varepsilon$$

for some small $\varepsilon > 0$, and since

$$\frac{1}{2}\omega \leq \omega_t' = \omega + \partial\bar{\partial}\psi_t \leq 2\omega,$$

the first of the above observations gives us the following contradiction

$$\|\psi_{t_1}\|_{C^{2,\frac{1}{2}}} \leq C\varepsilon < \frac{1}{4}.$$

This proves the claim. □

Remark 6.30 The norm $\|\psi_t\|_{C^{2,\frac{1}{2}}}$ might become bigger than $\frac{1}{4}$, but this can only happen for t outside $[t_0, 1]$.

For $t \geq \max\{t_0, 1 - \frac{1}{60}\}$, we get

$$\|\phi_t\|_{C^0} \geq (1 - 30(1-t))\|\phi_1\|_{C^0} - 1.$$

This in its turn implies that

$$\begin{aligned}
F_\omega(\phi) &= F_\omega(-\phi_1) \\
&= -F_{\omega'}(-\phi_1) \\
&= \int_0^1 (I_{\omega'}(\phi_t) - J_{\omega'}(\phi_t))dt \\
&\geq \int_{t_0}^1 (I_{\omega'}(\phi_t) - J_{\omega'}(\phi_t))dt \\
&\geq \min\left\{1 - t_0, \frac{1}{60}\right\}(I_{\omega'}(\phi_{t_0}) - J_{\omega'}(\phi_{t_0})) \\
&\geq \min\left\{1 - t_0, \frac{1}{60}\right\}(I_{\omega'}(\phi_1) - J_{\omega'}(\phi_1)) - C \\
&\geq \frac{1}{n}\min\left\{1 - t_0, \frac{1}{60}\right\} J_{\omega'}(\phi_1) - C.
\end{aligned}$$

Hence,

$$F_\omega(\phi) \geq \frac{1}{n}\min\left\{1 - t_0, \frac{1}{60}\right\} J_\omega(\phi) - C$$

and if $1 - t_0 \geq \frac{1}{60}$, this reads

$$F_\omega(\phi) \geq \frac{1}{60n} J_\omega(\phi) - C.$$

Thus, we are done. However, if $1 - t_0 < \frac{1}{60}$, we need to do more. We then have

$$\begin{aligned}
F_\omega(\phi) &\geq \frac{1 - t_0}{n} J_\omega(\phi) - C \\
&\geq \frac{c' J_\omega(\phi)}{(1 + \|\phi\|_{C^0})^{\frac{2n+2}{2n+2+\beta}}} - C,
\end{aligned}$$

where c' may depend on ε chosen in the first of the above two observations. If we have

$$\|\phi\|_{C^0} \leq C(1 + J_\omega(\phi)),$$

then it follows

$$F_\omega(\phi) \geq C' J_\omega(\phi)^{\frac{\beta}{2n+2+\beta}} - C.$$

This would prove the theorem, but we do not want the assumption on the C^0-norm of ϕ and in order to get rid of that, we need to do more work (due to X. Zhu and G. Tian).

Lemma 6.31 *There exists a constant $C > 0$ such that for $t \geq \frac{1}{2}$, we have*

$$\operatorname{osc}_M(\phi_t - \phi_1) \leq C(1 + J_\omega(\phi_t - \phi_1)),$$

where $\operatorname{osc}_M f = \sup_M f - \inf_M f$.

Proof. It suffices to prove the above estimate for I_ω instead of J_ω because we know how to bound I_ω in terms of J_ω. We know that

$$\omega^n = e^{h_{\omega_t} - (\phi_t - \phi_1)} \omega_t^n$$

and that

$$\operatorname{Ric}(\omega_t) \geq t\omega_t \geq \frac{1}{2}\omega_t, \ \operatorname{Ric}(\omega) = \omega,$$

which implies that we have uniform Sobolev constants for both ω_t and ω. We also have that

$$\omega + \partial\bar{\partial}(\phi_t - \phi_1) = \omega_t > 0,$$
$$\omega_t + \partial\bar{\partial}(\phi_1 - \phi_t) > 0,$$

which imply respectively that

$$\Delta_\omega(\phi_t - \phi_1) \geq -n,$$

$$\Delta_{\omega_t}(\phi_t - \phi_1) \leq -n.$$

Set $(\phi_t - \phi_1)_+ = \max\{\phi_t - \phi_1, 0\}$, then Moser's iteration method will provide us with the following bound

$$\sup_M (\phi_t - \phi_1)_+ \leq C\left(1 + \|(\phi_t - \phi_1)_+\|_{L^2}\right)$$

and we will obtain a similar estimate for $(\phi_t - \phi_1)_- = \min\{\phi_t - \phi_1, 0\}$. Using then the Poincaré inequality as before, we get the required bound by I_ω. □

Claim Setting $\omega_t = \omega' + \partial\bar{\partial}\phi_t$, we have

$$\frac{1}{V}\int_M \phi_t \omega_t^n \leq 0 \text{ and } J_{\omega'}(\phi_t) - \frac{1}{V}\int_M \phi_t \omega'^n \leq 0.$$

Proof. From (a), we have

$$\frac{d}{dt}\left(-t(I_{\omega'}(\phi_t) - J_{\omega'}(\phi_t)) - \frac{t}{V}\int_M \phi_t\omega_t^n\right) = -I_{\omega'}(\phi_t) + J_{\omega'}(\phi_t),$$

Integrating this as we did before, we see that

$$\frac{t}{V}\int_M \phi_t\omega_t^n = -t(I_{\omega'}(\phi_t) - J_{\omega'}(\phi_t)) + \int_0^t (I_{\omega'}(\phi_s) - J_{\omega'}(\phi_s))ds.$$

Since $I_{\omega'}(\phi_s) - J_{\omega'}(\phi_s)$ increases with s, the integral on the right is bounded above by

$$t(I_{\omega'}(\phi_t) - J_{\omega'}(\phi_t)).$$

Therefore, we have proved the first inequality claimed, namely,

$$\frac{1}{V}\int_M \phi_t\omega_t^n \leq 0.$$

By (b), we have

$$\frac{d}{dt}\left(t(J_{\omega'}(\phi_t) - \frac{1}{V}\int_M \phi_t\omega'^n)\right) = -(I_{\omega'}(\phi_t) - J_{\omega'}(\phi_t)).$$

This yields upon integrating from 0 to t

$$t\left(J_{\omega'}(\phi_t) - \frac{1}{V}\int_M \phi_t\omega'^n\right) \leq 0.$$

Since $t > 0$, this means

$$J_{\omega'}(\phi_t) - \frac{1}{V}\int_M \phi_t\omega'^n \leq 0.$$

The claim is proved. □

Notice that

$$\begin{aligned} F_{\omega'}(\phi_t) - F_{\omega'}(\phi_1) &= J_{\omega'}(\phi_t) - \frac{1}{V}\int_M \phi_t\omega'^n \\ &\quad - \log\left(\frac{1}{V}\int_M e^{-(1-t)\phi_t}\omega_t^n\right) - J_{\omega'}(\phi_1) + \frac{1}{V}\int_M \phi_1\omega'^n. \end{aligned}$$

Using the concavity of logarithm, we have

$$-\log\left(\frac{1}{V}\int_M e^{-(1-t)\phi_t}\omega_t^n\right) \leq \frac{1-t}{V}\int_M \phi_t\omega_t^n.$$

Hence, it follows from the claim

$$\begin{aligned}
&F_{\omega'}(\phi_t) - F_{\omega'}(\phi_1)\\
\leq & t\Big(J_{\omega'}(\phi_t) - \frac{1}{V}\int_M \phi_t\omega'^n\Big) - \Big(J_{\omega'}(\phi_1) - \frac{1}{V}\int_M \phi_1\omega'^n\Big)\\
= & -\int_t^1 \frac{d}{dt}\Big(t(J_{\omega'}(\phi_t) - \frac{1}{V}\int_M \phi_t\omega'^n)\Big)dt\\
= & \int_t^1 \Big(I_{\omega'}(\phi_t) - J_{\omega'}(\phi_t)\Big)dt.
\end{aligned}$$

Since $I_{\omega'}(\phi_t) - J_{\omega'}(\phi_t)$ is increasing, the above gives us the following

$$F_{\omega'}(\phi_t) - F_{\omega'}(\phi_1) \leq (1-t)\Big(I_{\omega'}(\phi_1) - J_{\omega'}(\phi_1)\Big).$$

This leads us to, using that $\phi_1 = -\phi$,

$$\begin{aligned}
(1-t)J_\omega(\phi) &= (1-t)\Big(I_{\omega'}(\phi_1) - J_{\omega'}(\phi_1)\Big)\\
&\geq F_{\omega'}(\phi_t) - F_{\omega'}(\phi_1)\\
&= F_\omega(\phi_t - \phi_1).
\end{aligned}$$

On the other hand, by the last lemma and what we have already proved, we get that

$$F_\omega(\phi_t - \phi_1) \geq c_1 \operatorname{osc}_M(\phi_t - \phi_1)^{\delta(n)} - c_2,$$

and consequently,

$$(1-t)J_\omega(\phi) \geq c_1 \operatorname{osc}_M(\phi_t - \phi_1)^{\delta(n)} - c_2,$$

therefore, for $t \geq t_0$, we finally get

$$\begin{aligned}
F_\omega(\phi) &\geq (1-t)(I_{\omega'}(\phi_t) - J_{\omega'}(\phi_t))\\
&\geq \frac{1-t}{n}J_{\omega'}(\phi_t)\\
&\geq \frac{1-t}{n}J_{\omega'}(\phi_1) - 2(1-t)\operatorname{osc}_M(\phi_t - \phi_1)\\
&\geq \frac{1-t}{n}J_\omega(\phi) - 2(1-t)c_1^{\frac{1}{\delta}}((1-t)J_\omega(\phi) + c_2)^{\frac{1}{\delta}}
\end{aligned}$$

and the theorem now follows by choosing $(1-t) \sim (1 + J_\omega(\phi))^{-1+\delta}$. □

6.3 Examples

In this section, we apply Theorem 6.10 to prove existence of Kähler-Einstein metrics on Fermat hypersurfaces in complex projective space. It was proved that any Fermat hypersurface of degree p admits a Kähler-Einstein metric in [25] in the case of $p \geq n$, and in [20] in the case of $n-1 \geq p \geq \frac{n+1}{2}$, where n is the complex dimension of the hypersurface. We will also discuss briefly the existence of Kähler-Einstein metrics on complex surfaces at the end of this section.

A Fermat hypersurface of degree p and complex dimension n is defined as the zero locus

$$M = \{[z_0 : \cdots : z_{n+1}] \in \mathbb{C}P^{n+1} \mid z_0^p + \cdots + z_{n+1}^p = 0\} \subset \mathbb{C}P^{n+1}.$$

From Example 2.9, we know that if $p < n+2$, then $c_1(M) > 0$. We will confine ourselves to the cases: $p < n+2$.

Consider the group G_0 generated by

$$\sigma_i : [z_0 : \cdots : z_{n+1}] \mapsto [z_0 : \cdots : e_p z_i : \cdots : z_{n+1}]$$

where $e_p = e^{\frac{2\pi\sqrt{-1}}{p}}$. Clearly M is invariant under G_0. Let G be the maximal compact subgroup of the automorphism group of $\mathrm{Aut}M$ containing G_0. Choose a G-invariant metric ω. We will show that F_ω is proper on $P_G(M,\omega)$. Let

$$\pi_i : M \to \mathbb{C}P_i^n$$

be the projection onto

$$\mathbb{C}P_i^n = \{[z_0 : \cdots : z_{i-1} : 0 : z_{i+1} : \cdots : z_{n+1}]\} \simeq \mathbb{C}P^n.$$

Note that this map is well defined because

$$[0 : \cdots : 0 : 1 : 0 : \cdots : 0] \notin M.$$

So any $\phi \in P_G(M,\omega)$ is of the form

$$(n+2-p)\pi_i^*\varphi$$

for some $\varphi \in P(M,\omega_{FS})$, where ω_{FS} is the Fubini-Study metric on $\mathbb{C}P^n$. Note that

$$c_1(M) = (n+2-p)[\pi^*\omega_{FS}].$$

We may normalize

$$\int_M e^{h_\omega - \phi}\omega^n = V.$$

By the cocycle property of F^0, we have

$$F_\omega(\phi) = F^0_\omega(\phi) = F^0_{(n+2-p)\pi_i^*\omega_{FS}}(\phi) - F^0_\omega(u),$$

where $(n+2-p)\pi_i^*\omega_{FS} = \omega + \partial\bar{\partial}u$ might not be a Kähler metric anymore, but u and therefore $F^0_\omega(u)$ are bounded. Using the basic properties of F^0_ω described in Section 6.1, we have

$$F^0_{(n+2-p)\pi_i^*\omega_{FS}}(\phi) = \frac{n+2-p}{n+1} F^0_{(n+1)\omega_{FS}}\left(\frac{n+1}{n+2-p}\varphi\right).$$

Since $\mathbb{C}P^n$ has a canonical Kähler-Einstein metric $(n+1)\omega_{FS}$, we know that $F_{(n+1)\omega_{FS}}$ is bounded from below on $P(\mathbb{C}P^n, (n+1)\omega_{FS})$. This implies that

$$F_\omega(\phi) \geq \frac{n+2-p}{n+1}\log\left(\frac{1}{V}\int_M e^{-\frac{n+1}{n+2-p}\phi}(n+2-p)^n \pi_i^*\omega_{FS}^n\right) - C,$$

where C always denotes a uniform constant. Consequently,

$$F_\omega(\phi) \geq \frac{n+2-p}{n+1}\log\left(\frac{1}{V}\int_M e^{-\frac{n+1}{n+2-p}\phi}(n+2-p)^n \sum_{i=0}^{n+1}\pi_i^*\omega_{FS}^n\right) - C,$$

since $\sum_{i=0}^{n+1}\pi_i^*\omega_{FS}^n \geq c\omega$ for some positive constant $c > 0$, we have

$$F_\omega(\phi) \geq \frac{n+2-p}{n+1}\log\left(\frac{1}{V}\int_M e^{-\frac{p-1}{n+2-p}\phi}e^{h_\omega-\phi}\omega^n\right) - C.$$

We may assume that $p > 1$.

Claim There is a uniform $C > 0$ such that

$$\sup_M \phi \leq C\left(1 + \log\left(\frac{1}{V}\int_M e^{-\frac{p-1}{n+2-p}\phi}e^{h_\omega-\phi}\omega^n\right)\right).$$

Proof. It was proved in [25] that there are $\alpha > 0$ and $C > 0$, such that for any $\phi \in P(M,\omega)$,

$$\frac{1}{V}\int_M e^{-\alpha(\phi - \sup_M \phi)}\omega^n \leq C.$$

Choose $\delta > 0$ such that

$$\alpha = \frac{\delta(n+1)}{\delta(n+2-p)+p-1}.$$

Then by the Sobolev inequality, we have

$$
\begin{aligned}
1= & \int_M e^{h_\omega-\phi}\omega^n \\
\leq & e^{-\delta\sup_M\phi}\left(\int_M e^{\alpha(h_\omega-\phi+\sup_M\phi)}\omega^n\right)^{\frac{\delta}{\alpha}}\left(\int_M e^{\frac{p-1}{n+2-p}(h_\omega-\phi)}e^{h_\omega-\phi}\omega^n\right)^{\frac{(1-\delta)(n+2-p)}{n+1}} \\
\leq & Ce^{-\delta\sup_M\phi}\left(\int_M e^{\frac{p-1}{n+2-p}(h_\omega-\phi)}e^{h_\omega-\phi}\omega^n\right)^{\frac{(1-\delta)(n+2-p)}{n+1}}.
\end{aligned}
$$

Then the claim follows. □

It follows from this claim that F_ω is proper on $P_G(M,\omega)$, and consequently, M admits a Kähler-Einstein metric.

We already saw that if M has a Kähler-Einstein metric, then the Calabi-Futaki invariant f_M vanishes. It was proved in [23] that the converse is also true for complex surfaces.

Theorem 6.32 *If $n=2$, then M has a Kähler-Einstein metric if and only if the Calabi-Futaki invariant $f_M=0$.*

To prove this, basically all one needs to do is check that F_ω is proper along the solutions of the corresponding Monge-Ampère equation. We refer the readers to [23] for its proof.

Theorem 6.10 can be also used to simplify the proof of this theorem for most complex surfaces by showing that F_ω is proper on $P_G(M,\omega)$. Let us illustrate it.

By the classification theory of complex surfaces, we know that any surface M in the above theorem is of the form either $\mathbb{C}P^1\times\mathbb{C}P^1$ or the blow-up of $\mathbb{C}P^2$ at k points ($0\leq k\leq 8$). Clearly, $\mathbb{C}P^1\times\mathbb{C}P^1$ and $\mathbb{C}P^2$ have homogeneous Kähler-Einstein metrics. We have shown before that the blow-up of $\mathbb{C}P^2$ at one or two points does not admit any Kähler-Einstein metrics, since they have non-vanishing Futaki invariants. It was shown in [26] that the blow-up of $\mathbb{C}P^2$ at $k=3$ or 4 points in general position has a Kähler-Einstein metric ($k=3$ was also proved by Siu [21]). Theorem 6.10 can be used to prove the existence of Kähler-Einstein metrics on any M with $c_1(M)>0$ which is a blow-up of $\mathbb{C}P^2$ at k points with $k=5,7,8$. The case that $k=6$ still needs to use the proof in [23] at this moment. As an example, let us show existence of Kähler-Einstein metrics on any blow-up M of $\mathbb{C}P^2$ at 7 points in general position (here $c_1(M)>0$).

Let M be a blow-up of $\mathbb{C}P^2$ at 7 points. Then it is a double covering

$$\pi: M\mapsto\mathbb{C}P^2$$

with branch locus along a smooth quartic curve. Assume that G contains the deck transformation of π. As before, ω denotes a fixed, G-invariant Kähler metric with the Kähler class $c_1(M)$. Then for any $\phi \in P_G(M,\omega)$, $\pi^*\pi_*\phi = 2\phi$. We normalize ϕ such that

$$\frac{1}{V}\int_M e^{h_\omega - \phi}\omega^2 = 1.$$

Put $f = \pi^*\omega_{FS}^2/\omega^2$. It is non-negative and vanishes along the branch locus of π. Moreover, we have

$$\frac{1}{V}\int_M |f|^{-\frac{4}{5}}\omega^2 \leq c < \infty.$$

Using the Hölder inequality, we deduce from this

$$\frac{1}{V}\int_M e^{-\frac{4}{3}\phi}\omega^2 \leq c\left(\frac{1}{V}\int_M e^{-3\phi}\pi^*\omega_{FS}^2\right)^{\frac{4}{9}}.$$

Note that c always denotes a uniform constant. By using Theorem 6.7 for $\mathbb{C}P^2$, we can show (compare the last example) that

$$F_\omega(\phi) \geq \frac{1}{3}\log\left(\frac{1}{V}\int_M e^{-3\phi}\pi^*\omega_{FS}^2\right) - c.$$

It follows from the above inequalities that

$$F_\omega(\varphi) \geq \frac{1}{4}\sup_M \varphi - c.$$

Therefore, F_ω is proper on $P_G(M,\omega)$, and consequently by Theorem 6.10, M admits a Kähler-Einstein metric.

Chapter 7

Applications and generalizations

In this chapter, we will discuss some applications of theorems in previous chapters. We will also give some generalizations of previous results.

7.1 A manifold without Kähler-Einstein metric

We will now consider the case $n \geq 3$ and here we will need to introduce some new concepts.

Definition 7.1 *A holomorphic degeneration of M is a fibration $\pi : W^{n+1} \to D$ such that $\pi^{-1}(\frac{1}{2}) \simeq M$ and $\pi^{-1}(t)$ is smooth for all $t \neq 0$. Here D is the unit disc in the complex plane*

Example 7.2 The easiest example of such a degeneration is the following: Let $W = \{(z_1, z_2) \in \mathbb{C}^2\}$ and define $\pi(z_1, z_2) = z_1 z_2$ then $\pi^{-1}(t) = \{z_1 z_2 = t\}$. It is clear that this is smooth for $t \neq 0$ but not smooth at $t = 0$.

Definition 7.3 *A special degeneration is a degeneration that in addition satisfies: there exists a holomorphic vector field v on W such that $\pi_* v = -t\frac{\partial}{\partial t}$ (generating a one-parameter subgroup $z \mapsto e^{-t}z$ on D).*

Remark 7.4 The vector field $v|_{\pi^{-1}(t)}$ is not necessarily tangent to $\pi^{-1}(t)$ when $t \neq 0$, but $v|_{\pi^{-1}(0)}$ is, because 0 is a fixed point. Note that all fibers $\pi^{-1}(t)$ except $\pi^{-1}(0)$ are biholomorphic to M in a special degeneration.

In this section, for simplicity, we always assume that the central fiber $\pi^{-1}(0)$ is smooth. We refer the readers to [24] for general cases where the central fiber may not be smooth.

Example 7.5 (Trivial product) Let $W = M \times D$ so that $v = v_1 - t\frac{\partial}{\partial t}$ for $v_1 \in \eta(M)$. Observe that $\pi^{-1}(t) \simeq \pi^{-1}(s)$ when t, s are both non-zero. This is true because v generates a complex one-parameter group of automorphisms ϕ_t of W such that $\pi \circ \phi_t(w) = e^{-t}\pi(w)$ for all $w \in W$. So all the fibers over D are isomorphic.

In general, $\pi^{-1}(0)$ might not be isomorphic to $\pi^{-1}(t)$ for $t \neq 0$ and we say in this case that the complex structure jumps at $\pi^{-1}(0)$.

Example 7.6 Let M_{p_4} be the blow up of $\mathbb{C}P^2$ in four points $p_1, \dots, p_4$, where $p_1 = [1:0:0], p_2 = [0:1:0]$ and $p_3 = [0:0:1]$ are fixed. If p_4 is not on any of the lines $\overline{p_1p_2}, \overline{p_1p_3}$ or $\overline{p_2p_3}$, then the complex structure of M_{p_4} is independent of p_4, since there is an automorphism of $\mathbb{C}P^2$ fixing p_1, p_2 and p_3 and bringing p_4 to $[1:1:1]$ in this case. Note that in this case, $c_1(M) > 0$, which is equivalent to the fact that there are no holomorphic $\mathbb{C}P^1$'s with self intersection number less than or equal to -2.

If, however, for example $p_4 \in \overline{p_2p_3}$, then the complex structure changes. To see this, let E be the line in M_{p_4} over $\overline{p_2p_3} \subset \mathbb{C}P^2$, that is we have blown up in three points on $\overline{p_2p_3}$ and each time we blow up, we reduce the intersection number by 1, so $E^2 = -2$. Set now

$$\pi : W = \bigcup_{t \in D} M_{p_4(t)} \to D$$

sending $M_{p_4(t)}$ to t. Assume $p_4(0) \in \overline{p_2(0)p_3(0)}$, then $t = 0$ corresponds to the jumping of complex structure.

Remark 7.7 There is no complex surface M with $c_1(M) > 0$ that admits a non-trivial special degeneration $\pi : W \to D$ such that $c_1(\pi^{-1}(0)) > 0$. To see this, use the classification theory of surfaces.

Theorem 7.8 *If M has a Kähler-Einstein metric with $c_1(M) > 0$, then for every special degeneration $\pi : W \to D$ we have that*

$$\mathrm{Re}(f_{\pi^{-1}(0)}(v|_{\pi^{-1}(0)})) \geq 0$$

with equality if and only if $W = M \times D$.

The following example was missed by Fano in his classification of Fano 3-folds (three-dimensional complex manifolds with positive first Chern class) and was found by Iskovskih. The author learned this description from S. Mukai (see [19]). It forms a counterexample to a folklore conjecture, which is none the less true in dimension two.

Conjecture 7.9 *One can always find a Kähler-Einstein metric on M if*

$$\eta(M) = \{0\}.$$

Example 7.10 We will give two descriptions of the following manifold. First of all, consider $Sl(2,\mathbb{C})/\Gamma$, where Γ is the icosahedral group. The manifold M we are interested in is the compactification of the above quotient. Let R_{12} be the space of homogeneous polynomials of degree 12 in two variables. Then

$$PR_{12} \simeq \mathbb{C}P^{12}.$$

Now $Sl(2,\mathbb{C})$ acts on R_{12} by composition and M is defined as

$$\overline{\text{Orbit of } SL(2,\mathbb{C})f},$$

where f is an appropriate, Γ-invariant polynomial in R_{12}.

The following description is perhaps easier to understand. Consider the Grassmannian $G(4,7)$ consisting of four-dimensional subspaces of $\mathbb{C}^7$. Take now any 3-dimensional subspace $P \subset \Lambda^2\mathbb{C}^7$ and define

$$X_P = \{E \in G(4,7); \pi_E(P) = 0\}$$

where $\pi_E : \Lambda^2\mathbb{C}^7 \to \Lambda^2 E^\perp$ is the orthogonal projection, then a dimension count tells us that we can expect that dim $X_P = 3$. We make the following two observations:

- If X_P is smooth and non-degenerate (that is no vector field $v \in \mathbb{C}^7$ vanishes in X_P), then $c_1(X_P) = c_1(Q)|_{X_P}$. Here Q is the universal quotient bundle on $G(4,7)$ and therefore $c_1(Q)|_{X_P} = \frac{1}{7}c_1(G(4,7))|_{X_P} > 0$.
- $\eta(M) = \{v \in sl(2,\mathbb{C});$ induced action by v preserves $P\}$, where

$$sl(2,\mathbb{C}) = \eta(G(4,7)).$$

Now let P_0 be the span of

$$\begin{aligned} u_1 &= 3e_1 \wedge e_6 - 5e_2 \wedge e_5 + 6e_3 \wedge e_4 \\ u_2 &= 3e_1 \wedge e_7 - 2e_2 \wedge e_6 + e_3 \wedge e_5 \\ u_3 &= e_2 \wedge e_7 - e_3 \wedge e_6 + e_4 \wedge e_5 \end{aligned}$$

which is invariant under the irreducible representation of $sl(2,\mathbb{C})$ on $\mathbb{C}^7$ given by

$$\begin{pmatrix} 1 & 0 \\ 0 & -1 \end{pmatrix} \mapsto \text{diag}\{3,2,1,0,-1,-2,-3\}$$

$$\begin{pmatrix} 0 & 0 \\ 1 & 0 \end{pmatrix} \mapsto \begin{pmatrix} 0&1&0&0&0&0&0 \\ 0&0&1&0&0&0&0 \\ 0&0&0&1&0&0&0 \\ 0&0&0&0&1&0&0 \\ 0&0&0&0&0&1&0 \\ 0&0&0&0&0&0&1 \\ 0&0&0&0&0&0&0 \end{pmatrix}$$

$$\begin{pmatrix} 0 & 1 \\ 0 & 0 \end{pmatrix} \mapsto \begin{pmatrix} 0 & 0 & 0 & 0 & 0 & 0 & 0 \\ 3 & 0 & 0 & 0 & 0 & 0 & 0 \\ 0 & 5 & 0 & 0 & 0 & 0 & 0 \\ 0 & 0 & 6 & 0 & 0 & 0 & 0 \\ 0 & 0 & 0 & 6 & 0 & 0 & 0 \\ 0 & 0 & 0 & 0 & 5 & 0 & 0 \\ 0 & 0 & 0 & 0 & 0 & 3 & 0 \end{pmatrix}.$$

Check that X_{P_0} is smooth and non-degenerate. This implies that $\eta(X_{P_0}) = sl(2,\mathbb{C})$ is semi-simple and by Corollary 3.6, we know that this implies that $f_{X_{P_0}} = 0$. We will now deform P_0 to

$$P_a = \operatorname{span}\{u_i + \sum_{j+k \geq i+7} a_{ijk} e_j \wedge e_k, i = 1,2,3\}$$

then $\eta(X_{P_a}) = \{0\}$ for generic a and therefore by the above observation, we have

$$\sigma(t)(P_a) \to P_0, \text{ when } \operatorname{Re}(t) \to \infty,$$

where $\sigma(t)$ denotes the diagonal matrix

$$\operatorname{diag}(e^{3t}, e^{2t}, e^t, 1, e^{-t}, e^{-2t}, e^{-3t})$$

for $t \in \mathbb{C}$. This follows that

$$X_{\sigma(t)P_a} \simeq X_{P_a} \text{ and } X_{\sigma(t)P_a} \to X_{P_0}.$$

So set

$$\pi : W = \Big(\bigcup_{\operatorname{Re}(t) \geq 0} X_{\sigma(t)P_a} \Big) \cup X_{P_0} \to D$$

sending $X_{\sigma(t)P_a}$ to e^{-t}. So we have here a non-trivial degeneration (since $X_{P_a} \neq X_{P_0}$) and the Calabi-Futaki invariant vanishes ($f_{X_{P_0}} = 0$) so in this case, Theorem 7.8 implies that X_{P_a} has no Kähler-Einstein metric.

We will not prove Theorem 7.8 in this section (see for a proof [24]), but we will explain briefly some ideas for the proof. Fix any Kähler metric Ω on W satisfying $[\Omega|_{\pi^{-1}(t)}] = c_1(\pi^{-1}(t))$ and let Φ_t be a biholomorphic map from M onto $\pi^{-1}(t)$ for $t \neq 0$ generated by the vector field v on W. If we set

$$\omega_t = \Omega|_{\pi^{-1}(t)},$$

then one can check that $\Phi_t^* \omega_t$ is in the class of $c_1(M)$, so

$$\Phi_t^* \omega_t = \omega + \partial\bar{\partial}\phi_t.$$

One can show

$$\lim_{t \to 0} \frac{d}{dt} F_\omega(\phi_t) = \operatorname{Re}(f_{\pi^{-1}(0)}(v)).$$

Because F_ω is proper, this proves the theorem. To get the equality part, one needs to know more about the rate of convergence when $t \to 0$.

7.2 K-energy and metrics of constant scalar curvature

In Chapter 4, we learned that the scalar curvature of Kähler metrics is a moment map for the action of symplectic diffeomorphisms on the space of almost complex structures compatible with a given symplectic form. In this section, we will introduce the K-energy, which was first defined by Mabuchi in [18], and discuss its basic properties.

Let (M,ω) be a compact Kähler manifold. As before, we denote by $P(M,\omega)$ the space

$$\{\phi \in C^\infty; \omega + \partial\bar\partial\phi > 0\}.$$

For $\phi \in P(M,\omega)$, we define its K-energy by

$$\nu_\omega(\phi) = -\frac{1}{V}\int_0^1\int_M \dot\phi_t(s(\omega_t) - n\mu)\omega_t^n \wedge dt,$$

where $\{\phi_t\}$ is any path in $P(M,\omega)$ with $\phi_0 = 0$ and

$$\phi_1 = \phi, \omega_t = \omega + \partial\bar\partial\phi_t,$$

and $s(\omega_t)$ is its scalar curvature and

$$\mu = \frac{c_1(M)\cdot[\omega]^{n-1}}{[\omega_t]^n}.$$

Of course, in order for $\nu_\omega(\phi)$ to be well defined, we need first to make sure that the integral on the right is independent of paths connecting 0 to ϕ. In fact, integrating by parts, one can show the integral on the right is equal to

$$\frac{1}{V}\int_M \log\left(\frac{\omega_\phi^n}{\omega^n}\right)\omega_\phi^n$$
$$-\sum_{i=0}^{n-1}\frac{1}{V}\int_M \phi\,\mathrm{Ric}(\omega)\wedge\omega^i\wedge\omega_\phi^{n-1-i} + \frac{n\mu}{n+1}\sum_{i=0}^{n}\frac{1}{V}\int_M \phi\omega^i\wedge\omega_\phi^{n-i}.$$

It follows that ν_ω is well defined.

From the definition, one can easily show that ν_ω satisfies the cocycle condition as F_ω does, more precisely, if ω' is another Kähler metric of the form

$$\omega + \partial\bar\partial\psi,$$

then

$$\nu_\omega(\phi) - \nu_{\omega'}(\phi - \psi) = \nu_\omega(\psi).$$

If ω is a Kähler metric with constant scalar curvature, then the K-energy ν_ω can be expanded as

$$\nu_\omega(t\phi) = \frac{t^2}{V}\int_M \phi_{ij}\overline{\phi}_{ij}\omega^n + O(t^3).$$

It follows that any Kähler metric of constant scalar curvature is a minimum of the K-energy, moreover, it is a strict minimum if there are no functions ϕ with $\phi_{ij} = 0$. Note that this last condition is satisfied if and only if $\eta(M) = \{0\}$.

Now we let X be a holomorphic vector field and Φ_t be the integral curve of its real part $\mathrm{Re}(X)$. Notice that $\Phi_t^*\omega$ has the same Kähler class $[\omega]$, so we have ϕ_t such that $\Phi_t^*\omega = \omega + \partial\bar{\partial}\phi_t$. Differentiating this, we get

$$d(i_{\mathrm{Re}(X)}\omega) = \partial\bar{\partial}\dot{\phi}_t.$$

This implies that $i_{\mathrm{Re}(X)}\omega = \bar{\partial}\dot{\phi}_t + \alpha$, where α is a harmonic $(0,1)$-form. From the definition of the K-energy, we can then show

$$\frac{d}{dt}\nu_\omega(\phi_t) = -\frac{1}{V}\int_M \dot{\phi}_t(s(\omega_{\phi_t}) - n\mu)\omega_{\phi_t}^n = \frac{1}{V}\mathrm{Re}(f_M([\omega], \mathrm{Re}(X))).$$

In particular, if the Calabi-Futaki invariant is non-zero, the K-energy ν_ω is not bounded from below.

We have seen that the K-energy shares many properties of F_ω in the case of Kähler-Einstein metrics. However, we expect more.

Definition 7.11 *We say that ν_ω is proper on a closed subset $E \subset P(M,\omega)$ if there is an non-decreasing function $\lambda(t)$ with $\lim_{t\to\infty}\lambda(t) = \infty$, such that $\nu_\omega(\phi) \geq \lambda(J_\omega(\phi))$ for any $\phi \in E$, where J_ω is defined as before.*

As for the functional F_ω, the properness of ν_ω is independent of initial metric ω in an appropriate sense.

In view of the previous results in the last two chapters, it is reasonable to expect that the following is true.

Conjecture 7.12

- *If M has a Kähler metric ω of constant scalar curvature, then $\nu_\omega \geq 0$, particularly, it is bounded from below;*
- *If $\eta(M) = \{0\}$, then M has a Kähler metric of constant scalar curvature and Kähler class $[\omega]$ if and only if ν_ω is proper over $P(M,\omega)$;*
- *Let G be a maximal compact subgroup in $\mathrm{Aut}(M)$. Then M has a Kähler metric of constant scalar curvature and Kähler class $[\omega]$ if and only if ν_ω is proper over $P_G(M,\omega)$, where $P_G(M,\omega)$ consists of G-invariant functions in $P(M,\omega)$.*

Theorem 7.13 *Let (M,ω) be a compact Kähler manifold. If either $[\omega] = \pm c_1(M)$ or $c_1(M) = 0$, then the above conjecture is true.*

Proof. This theorem is essentially contained in results of the last two chapters. For the readers' convenience, we outline its proof here.

Under our assumptions, we have $\mathrm{Ric}(\omega) - \mu\omega = \partial\bar\partial h_\omega$ for some smooth function h_ω. Then by a straightforward computation, we get

$$\begin{aligned}\nu_\omega(\phi) =& \frac{1}{V}\int_M \log\left(\frac{\omega_\phi^n}{\omega^n}\right)\omega_\phi^n + \frac{1}{V}\int_M h_\omega(\omega^n - \omega_\phi^n) \\ &+ \frac{n\mu}{n+1}\frac{1}{V}\int_M \phi\omega^i \wedge \omega_\phi^{n-i} - \frac{\mu}{n+1}\sum_{i=1}^{n}\frac{1}{V}\int_M \phi\omega^i\wedge\omega_\phi^{n-i} \\ =& \frac{1}{V}\int_M \log\left(\frac{\omega_\phi^n}{\omega^n}\right)\omega_\phi^n + \frac{1}{V}\int_M h_\omega(\omega^n - \omega_\phi^n) - \mu(I_\omega(\phi) - J_\omega(\phi)).\end{aligned}$$

Here we have used

$$I_\omega(\phi) - J_\omega(\phi) = \sum_{i=0}^{n-1}\frac{n-i}{n+1}\frac{1}{V}\int_M \phi\wedge\omega^i\wedge\omega_\phi^{n-i}.$$

It was shown in [25] that there is an $\alpha(M,\omega) > 0$, depending only on M and ω, such that for any $\phi \in P(M,\omega)$,

$$\frac{1}{V}\int_M e^{-\log\left(\frac{\omega_\phi^n}{\omega^n}\right) - \beta(\phi - \sup_M \phi)}\omega_\phi^n = \frac{1}{V}\int_M e^{-\beta(\phi-\sup_M\phi)}\omega^n \leq C,$$

where β is any positive number which is less than $\alpha(M,\omega)$. By the convexity of the exponential function, we deduce from this

$$\frac{1}{V}\int_M \log\left(\frac{\omega_\phi^n}{\omega^n}\right)\omega_\phi^n \geq -\frac{\beta}{V}\int_M (\phi - \sup_M \phi)\omega_\phi^n - C.$$

Combining this with the above inequality of ν_ω, we get

$$\nu_\omega(\phi) \geq \beta I_\omega(\phi) - \mu(I_\omega(\phi) - J_\omega(\phi)) - C.$$

This implies that ν_ω is proper on $P(M,\omega)$ when $\mu \leq 0$. In fact, even if $\mu = 1$, ν_ω is still proper whenever $\alpha(M,\omega) > \frac{n}{n+1}$. Since M has a Kähler-Einstein metric when $c_1(M) \leq 0$ (compare with Chapter 5), we have proved this theorem in the cases that $\mu \leq 0$.

Now we assume that $\mu = 1$. It was proved in [7] that

$$F_\omega(\phi) = \nu_\omega(\phi) + \frac{1}{V}\int_M h_{\omega_\phi}\omega_\phi^n - \frac{1}{V}\int_M h_\omega\omega^n.$$

Since

$$\frac{1}{V}\int_M e^{h_{\omega_\phi}}\omega_\phi^n = 1, \qquad \text{we have} \qquad \frac{1}{V}\int_M h_{\omega_\phi}\omega_\phi^n \leq 0.$$

Then this theorem can be easily deduced from Theorem 6.7 and 6.10. □

Remark 7.14 If we assume that ω is a Kähler-Einstein metric with

$$Ric(\omega) = \mu\omega,$$

then the arguments in the above proof yield

$$\nu_\omega(\phi) = \frac{1}{V}\int_M \log\left(\frac{\omega_\phi^n}{\omega^n}\right)\omega_\phi^n - \mu\Big(I_\omega(\phi) - J_\omega(\phi)\Big).$$

This implies that if $\mu \leq 0$, then

$$\nu_\omega(\phi) \geq 0,$$

and the equality holds if and only if $\phi = 0$ constant. Therefore, the Kähler-Einstein metric ω attains the absolute minimum of the K-energy. The same is true even if $\mu = 1$. This was proved by Bando and Mabuchi.

7.3 Relation to stability

Let M be an algebraic manifold embedded in a complex projective space $\mathbb{C}P^N$. Let H be the hyperplane line bundle over $\mathbb{C}P^N$. This bundle gives rise to a Kähler class $[\omega]$, which is a positive multiple of $c_1(H|_M)$.

In this section, we relate the properness of the K-energy ν_ω to the stability of the underlying polarized manifold (M, ω). This is inspired by a conjecture of Yau on existence of Kähler-Einstein metrics with positive scalar curvature.

The stability of (M, ω) is described in terms of the following algebraic family.

Let $G = SL(r, \mathbb{C})$ $(r \geq 2)$. Let $\pi : \mathcal{X} \mapsto Z$ be a G-equivariant holomorphic fibration between smooth varieties, such that

(1) all fibers are connected subvarieties of complex dimension n and M is one of them;

(2) there is a G-equivariant embedding of $\mathcal{X}$ into $Z \times \mathbb{C}P^N$ for some N. Write $L = \pi_2^* H$, where H is the hyperplane bundle over $\mathbb{C}P^N$. Furthermore, $c_1(L|_M)$ is a positive multiple of $[\omega]$.

Clearly, L is a G-equivariant line bundle over $\mathcal{X}$, which is relatively ample over Z and has fixed topological type along smooth fibers. This last property can be used to restate (2) in a more intrinsic way.

Consider the virtual bundle

$$\mathcal{E} = (n+1)(\mathcal{K}^{-1} - \mathcal{K}) \otimes (L - L^{-1})^n - n\mu(L - L^{-1})^{n+1},$$

where $\mathcal{K} = K_{\mathcal{X}} \otimes K_Z^{-1}$ is the relative canonical bundle, and as before,

$$\mu = \frac{c_1(M) \cdot c_1(L|_M)^{n-1}}{c_1(L|_M)^n}.$$

We define L_Z to be the inverse of the determinant line bundle $\det(\mathcal{E}, \pi)$. By the Grothendieck-Riemann-Roch Theorem, we can compute

$$c_1(L_Z) = 2^{n+1}\pi_{1*}\left((n+1)c_1(\mathcal{K})c_1(L)^n + n\mu c_1(L)^{n+1}\right).$$

We also denote by $\mathcal{L}_Z^{-1}$ the total space of the line bundle L_Z^{-1} over Z. Then G acts naturally on $\mathcal{L}_Z^{-1}$. Recall that $X_z = \pi^{-1}(z)$ $(z \in Z)$ is weakly Mumford stable with respect to L, if the orbit $G \cdot \tilde{z}$ in $\mathcal{L}_Z^{-1}$ is closed, where $\tilde{z}$ is any non-zero vector in the fiber of L_Z^{-1} over z; if, in addition, the stabilizer G_z of z is finite, then X_z is Mumford stable. We also recall that X_z is Mumford semistable, if the 0-section is not in the closure of $G \cdot \tilde{z}$. Clearly, this stability (resp. semistability) is independent of choices of $\tilde{z}$.

Theorem 7.15 *Let $\pi : \mathcal{X} \mapsto Z$ be as above. Assume that $M = X_z$ and ν_ω is proper on $P(M, \omega)$. Then M is Mumford stable with respect to L.*

This theorem was proved in Section 8 of [24] without explicitly stating it. We refer the readers to [24] for its proof. The arguments in Section 8 of [24] also shows that the converse to the statement of Theorem 7.14 is true under slightly stronger assumption on the family $\pi : \mathcal{X} \mapsto Z$, namely the fibers have no multiple components.

Remark 7.16 More generally, if $\eta(M)$ is non-trivial, we may define the weak properness of ν_ω, which means, roughly speaking, properness of ν_ω modulo action of $\mathrm{Aut}_0(M)$, the identity component of $\mathrm{Aut}(M)$. This weak properness will imply the weakly Mumford stability. However, we will not discuss this general case here.

Now we discuss implications of the above theorem for the existence of Kähler-Einstein metrics with positive scalar curvature. Assume that (M, ω) be a compact Kähler manifold with $[\omega] = c_1(M)$. In late 80's, Yau proposed

Conjecture 7.17 *M admits a Kähler-Einstein metric if and only if it satisfies a certain stability condition in the sense of Mumford.*

Combining the above theorem with the theorem of last section, we have

Theorem 7.18 *[24] Let (M, ω) be and $\pi : \mathcal{X} \mapsto Z$ be as above. Assume that $M = X_z$ admits a Kähler-Einstein metric of positive scalar curvature. Then X_z is weakly Mumford stable. If X_z has no non-trivial holomorphic vector fields, then ν_ω is proper on $P(M, \omega)$ and M is actually Mumford stable with respect to L.*

This answers Yau's conjecture partially.

Example 7.19 Let us apply the above theorem to proving again non-existence of Kähler-Einstein metrics in Example 7.10. We will adopt the notations in the previous sections.

Recall that $W = G(4,7)$ consists of all 4-subspaces in $\mathbb{C}^7$. Let Q be its universal quotient bundle.

Let π_i $(i = 1,2)$ be the projection from $W \times G(3, H^0(W, \wedge^2 Q))$ onto its ith-factor, and let S be the universal bundle over $G(3, H^0(W, \wedge^2 Q))$.

We define

$$\mathcal{X} = \{(x, P) \in W \times G(3, H^0(W, \wedge^2 Q))\}$$

One can show that $\mathcal{X}$ is smooth.

If $L = \det(Q)$, then $c_1(L)$ is the positive generator of $H^2(W, \mathbb{Z})$.

Consider the fibration $\pi = \pi_2|_{\mathcal{X}} : \mathcal{X} \mapsto Z$, where

$$Z = \{P \in G(3, H^0(W, \wedge^2 Q)) \mid \dim_{\mathbb{C}} X_P = 3\}.$$

Its generic fibers are smooth Fano 3-folds.

Using the Adjunction Formula, one can show

$$c_1(\mathcal{K}) \;=\; -\pi_1^* c_1(L) - 3\pi_2^* c_1(S).$$

Therefore, it follows that

$$c_1(L_Z) = 16\pi_* \left(12\pi_2^* c_1(S^*)\pi_1^* c_1(L)^3 \;-\; \pi_1^* c_1(L)^4\right).$$

One can show that L_Z is ample.

By the definition of P_a, one can show that none of $G \cdot P_a$ is closed in $\mathcal{L}_Z^{-1}$. Therefore, none of generic X_{P_a} admits Kähler-Einstein metrics.

Bibliography

[1] T. Aubin *Nonlinear Analysis on Manifolds. Monge-Ampère Equations*, Springer-Verlag, Berlin-New York, 1982.

[2] R. Bott and S.S. Chern *Hermitian vector bundles and the equidistribution of the zeroes of their holomorphic sections*, Acta Math., 114, 1965, 71–112.

[3] A.L. Besse *Einstein manifolds*, Ergebnisse der Mathematik und ihrer Grenzgebiete; 3. Folge, Bd. 10, Springer-Verlag, 1987.

[4] E. Calabi *Extremal Kähler metrics*, Seminar on Diff. Geom., edited by Yau, Ann. of Math. Stud., 102, Princeton Univ. Press, 1982.

[5] J. Cheeger and D. Ebin *Comparison theorems in Riemannian geometry*, North-Holland Pub. Co., 1975.

[6] S.K. Donaldson *Remarks on Gauge theory, Complex Geometry and 4-manifold Topology*, Fields Medallists' lectures, 384–403, World Sci. Ser. 20th Century Math., 5, World Sci. Publishing.

[7] W. Ding and G. Tian *Kähler-Einstein metrics and the generalized Futaki invariants*, Invent. Math., 110 (1992), 315–335.

[8] A. Fijiki and G. Schumacher *The moduli space of Kähler structures on a real compact symplectic manifold*, Publ. Res. Inst. Math. Sci, 24, 1988, 141–168.

[9] A. Futaki *An obstruction to the existence of Einstein-Kähler metrics*, Inv. Math., 73 (1983), 437–443.

[10] A. Futaki *Kähler-Einstein Metrics and Integral Invariants*, Lecture Notes in Mathematics, 1314, Springer-Verlag.

[11] P. Griffiths & J. Harris *Principles of algebraic geometry*, Wiley Classics Library, New York, 1994.

[12] D. Gilbarg & N.S. Trudinger *Elliptic partial differential equations of second order*, Grundlehren der Mathematischen Wissenschaften, Springer-Verlag, Berlin-New York, 1983.

[13] S. Kobayashi, *On compact Kähler manifolds with positive definite Ricci tensor*, Annals of Mathematics 74, 1961.

[14] S. Kobayashi & K. Nomizu, *Differential Geometry*, Wiley-Interscience New York, 1963.

[15] S. Kobayashi & T. Ochiai, *Characterizations of complex projective spaces and hyperquadrics*, J. Math. Kyoto Univ. 13, 1973.

[16] M. Kontsevich, *Mirror symmetry in dimension 3*, Seminar Bourbaki, Vol. 94/95 No. 237, 1996 (page 286).

[17] P. Li, *On the Sobolev constant and the p-spectrum of a compact Riemannian manifold*, Ann. Sc. E.N.S. Paris 13, 1980.

[18] T. Mabuchi, *K-energy maps integrating Futaki invariants*, Tohoku Math. J., 38, 245–257 (1986).

[19] S. Mukai & H. Umemura, *Minimal rational threefolds*, Lecture notes in Math. 1016, Springer, Berlin-New York, 1983.

[20] A. Nadel *Multiplier ideal sheaves and existence of Kähler-Einstein metrics of positive scalar curvature*, Proc. Natl. Acad. Sci. USA, Vol. 86, No. 19, 1989.

[21] Y.T. Siu, *The Existence of Kähler-Einstein Metrics on Manifolds with Positive Anticanonical Line Bundle and a Suitable Finite Symmetry Group*, Ann. Math., 127, 1988, 585–627.

[22] G. Tian, *Kähler-Einstein metrics on algebraic manifolds*, Lecture notes in Math. 1646, Springer, Berlin-New York, 1997.

[23] G. Tian *On Calabi's conjecture for complex surfaces with positive first Chern class*, Inv. Math. Vol. 101, No. 1, 1990, 101–172.

[24] G. Tian, *Kähler-Einstein metrics with positive scalar curvature*, Inv. Math., 130 (1997), 1–39.

[25] G. Tian *On Kähler-Einstein metrics on certain Kähler Manifolds with* $C_1(M) > 0$, Invent. Math., 89 (1987), 225–246.

[26] G. Tian, G *Kähler-Einstein metrics on complex surfaces with* $C_1(M)$ *positive*, Comm. Math. Phy., 112, 1987.

[27] S.T. Yau, *On the Ricci curvature of a compact Kähler manifold and the complex Monge-Ampère equation I*, Comm. Pure Appl. Math., 31, 1978.

[28] S.T. Yau *On Calabi's conjecture and some new results in algebraic geometry*, Proc. Nat. Acad. Sci. USA, 74, 1977, 1798–1799.

Index